RAILWAY BEYOND

WATER MEADOWS

WALNUT TREE

NORTH PADDOCK

CUPRESSUS HE

HOME FIELD

THE FARM COTTAGE.

LOOKING WESTWARD ACROSS THE TEST VALLEY

A Plank Bridge by a Pool

NORMAN THELWELL
A Plank Bridge by a Pool

CHARLES SCRIBNER'S SONS • NEW YORK

Copyright © 1978 by Norman Thelwell

Library of Congress Cataloging in Publication Data

Thelwell, Norman, 1923-
  A plank bridge by a pool.

  1. Nature 2. Country life. 3. Fish ponds. 4. Fishing.
I. Title.
QH81.T34   500.9′422′732   78-13780
ISBN 0-684-15294-4

1 3 5 7 9 11 13 15 17 19 H/C 20 18 16 14 12 10 8 6 4 2
Printed in the United States of America

Thelwall [. . . *Thelewell* 1241 . . .]
'Pool by a plank bridge' . . .
The second el is OE waēl
'a weel, a deep pool, a deep
still part of a river.'

The Oxford Dictionary
of English Place-Names

Contents

Raby Mere

The Impossible Dream

Among the greatest pleasures of my childhood were the rare occasions when we were taken onto Raby Mere in a rowing boat. We lived at Birkenhead in the industrial north-west of England and this little lake, in its pretty country setting, was the nearest stretch of inland water. It was a favourite haunt for day trippers and local excursions but, in spite of its popularity, it retained a magnetic charm that impressed me deeply.

A cluster of rowing boats – long and narrow and with a touch of Victorian elegance – lined the small landing-stage where a man sat in a green and white striped hut and dispensed oars and warnings to the merry trippers who paid their money and clambered unsteadily into the boats. I remember the excitement of being lifted into one of them by strong adult hands, placed firmly on a seat (it had a cushion, what luxury) and warned by a big finger very near to my nose that I must not move.

Instructions were numerous and enforced with the rigour of army discipline. 'Don't fidget. Don't lean over the side. Don't stand up and *don't* keep grabbing at the oars or you'll have us all in the water.'

[9]

All good advice, of course, but a bit restricting to the spirit of a child who has just discovered the sheer delight of floating on the surface of this sparkling, dreamy element. No restrictions, however, could possibly destroy the new awareness of firm support below the boat, of the gentle rocking motion and the enchanting sound of the quicksilver ripple against the bow and along the smooth swell of the polished wooden sides.

Few experiences can match the heady pleasure of trailing one's hand gently through the cool smoothness of water, of feeling the surging movement, the gentle increase of pressure and caress between the fingers with each pull on the oars.

Sitting in a small boat on water has the same effect on the imagination as lying on one's stomach in the deep green of a summer meadow and watching the busy life of the sunlit jungle. One is so near to the water surface that one becomes part of it. Ducks float by almost on eye level like toys in the bath. A million tiny bits drift along a foot or two from the eye if one leans over the stern; each speck is animal or vegetable, each minute scrap has come from somewhere, has lived and grown and now, alive or dead, is going somewhere in the endless re-cycling process of time and space. Dimensions change completely. A small pool becomes as big and awesome and as wildly beautiful as Hickling Broad or Windermere. Every movement of the water by fish or wildfowl, by wind or rain is seen with new vision.

When I was old enough to cycle to Raby Mere with my friends and scrape a few pennies together we enjoyed the delicious luxury of taking out a boat alone, of learning how to use the oars without falling backwards off the seat and eventually savouring the thrill of feathering smoothly across the glassy surface with a minimum of splash and then drifting slowly round the curve into the quiet private pool at the far end of the lake, where a little stream ran gurgling into the deep, dark water.

I thought the Great God Pan lived there among the reeds and sedges and I rested on the oars and waited for the notes of his pipe and sometimes I thought I heard them.

There is no restriction, thank goodness, to dreams and the dreams of childhood are untouched by the tired dust of experience. I dreamed of having a lake of my own – deep, quiet and beautiful; where wild flowers grew on the banks and no ice-cream cartons and tiny wooden spoons floated at the lapping water's edge. The reeds would rustle in the wind and wild duck would sit motionless on a clutch of greenish-white eggs in the rank grass beneath the alders. I would drift slowly by in my own boat and see them there in the secret places pressed close to the ground and noticeable only by the tiny jewel of light in their patient eye. The thought of owning a boat and being free to push off from the bank without paying and

drifting idly about for hours and hours with no fear of an adult voice calling one back to reality was heady stuff. It floated me through childhood, and most tragedies and miseries sank below the water into oblivion and left the sun shining on an endless enchanted summer.

It was an impossible dream, of course, but that didn't matter. It was as legitimate as dreaming about driving a great steam engine or owning a Rolls Royce, or becoming King and letting everyone out of school forever.

Although I was a town child, or perhaps because of it, the countryside, which in those days could be reached easily by a penny bus ride almost from the front door, was like a glimpse of paradise and I really believed that heaven would be an endless sunlit landscape of woods and meadows and rippling water. I can remember the sharp thrill of fear that ran through me when I understood the Sunday school teacher to tell us that God had many mansions. I wanted to ask 'How many mansions? Is heaven a built-up area then, like Birkenhead?' But I hesitated to ask such awful questions: it might mean more black marks in the great book, perhaps, and I already had enough of those to make my acceptance into the Elysian Fields a doubtful prospect. So I let it go. I had never heard the word 'conservation' but deep down I felt sure that God would not cover the grass and wild flowers of heaven with bricks and mortar and hard grey paving stones, or let holy smoke besmirch the angels' wings like my mother's washing on Monday mornings.

I was reassured also by the fact that every year the Sunday school treats were a joyous trip to Raby Mere. Surely heaven would be the biggest treat of all, so there was bound to be a lake and ducks and free rides on the boats.

The mere was overlooked by two beautiful white cottages patterned with black half-timbering. Their roofs were thick reed thatch, scalloped and decorated like slabs of cake and each was set in its garden with flowerbeds and hanging baskets of geraniums and blue and white lobelia and alyssum. The green lawns were dotted with daisies and little rustic tables and chairs, where those with enough money for such luxuries could have tea and cakes brought to them in the sunshine and watch the lake whilst the teacups tinkled.

I do not remember envying these fortunate people, however, for we had cakes and buns and lemonade at trestle tables in the long low sheds by the swingboats in the field behind one of the cottages. The races were held there too: three-legged races, sack races and just plain races on the most undulating pasture I ever remember.

The field was so uneven that few of us children ever reached the tape held by shouting adults at the far end.

I won the only athletics prize of my life there, presented to me by a lady who picked me up somewhere along the race track, dusted my blouse and knees with her handkerchief and gave me a triangular cardboard tube of Toblerone chocolate. It seemed pointless to complete the course after that triumph so I wandered off to find my mother and have my head patted for winning something.

There seemed always to be lots of people about. A lady in a black straw hat with yellow roses on it and a shawl about her shoulders sat on the grassy bank below the tea-garden gate and sold pink rock from a big wicker basket. Ice-cream and sweets were obtainable from each cottage and even then, before the advent of plastic litter, toffee papers, cigarette packets and dark blue and white check cardboard wrappers from Walls ice-cream competed with the flowers in the grass.

Although the fact that it was difficult to be alone there irked me very much when I was a bit older, it did not seem to matter when we were young. There were fascinating ornate slot-machines by the tea sheds where we used to wait in little groups for people who had pennies to spend on such things. When these profligates turned up and fed the well-worn slots we would gather round to see the little iron horses race inside the glass case or the little cricketer (in a real knitted pullover – how marvellous!) jerk his bat madly at the marble which was bowled by the similarly clad figure at the other end. I don't remember seeing the ball struck once (what *do* you expect for a penny?) but it was exciting beyond words. The lordly young men who put their pennies in these machines were quite happy to let us watch but no amount of pleading would persuade them to let us have a peep into the machine called 'What the Butler Saw' once their penny had released the handle. When they had strolled away laughing and pushing one another we would lift each other up to peep through the metal eye-guard which kept the light off the two glass peep-holes. Our foreheads were bumped and grazed by the metal edge, for our friends never held us very steadily, but the secret within was always the same: a little square card which said 'Insert one penny in slot and turn handle'. We struggled with all our might but the handle would never turn. Children are very suspicious. Everyone insisted on trying before we wandered on into the little grassy dell below the lake.

The overflow stream tumbled down the side of this shady hollow and ran along the lower tree-lined fence before disappearing under the small stone road bridge. 'Jackies' were here in large numbers waving their tiny fins on the sandy bottom of the sun dappled stream. In all my childhood I never heard them called sticklebacks but only 'jack-sharps' or more often 'jackies'. The males in mating red were called 'doctors' – I wonder why? I once took about half-a-dozen of these fish on the bus ride home in an old H.P. Sauce bottle which I found in the sandy stream. I shed bitter tears when they were pronounced 'dead on arrival'. Who would expect fish to need air?

All these pleasures were optional extras, however, to the real thing, to the nucleus of it all: the shining mere with its boats and the more-than-happy white ducks that need never leave its water. I used to imagine myself one of those ducks on Monday mornings. How they must hug themselves with the warm arms of privilege as they floated serenely on the wide mirror or rested their heads over their snow-white backs and dozed away the gently rocking hours whilst men and women worked and children quaked through the drudging hours of school.

But there it was in this shining paradise that my unbelieving eyes saw the first shocking sight of violence and death. Two noble swans cruised on the lake with their downy grey cygnets, while the wild mallards and white ducks were attended by lines of tiny ducklings dashing here and there, getting lost and running along the surface to rejoin their brothers and sisters in a flurry of droplets.

The cob's neck arched and grew thicker and, breasting the water with a sudden surge, he reached out and seized a little duckling and smashed it again and again on the shattered glass surface, then held it under water for long dreadful seconds whilst my shout of protest stayed locked in my dry throat. The scrap of warm down floated back to the surface, a tiny webbed foot in the air, and revolved slowly in the eddy of the swan's wake.

The sun shone on and the water glittered. The swans cruised in merciless perfection and the ducks, it seemed, had noticed nothing. Three times more I watched the frightful destruction of a tiny living creature before I turned away with trembling stomach and knew, even then, that I had learned a lesson more important by far than my three-times table. As I grew older I came to realise that nature is cruel and indifferent to man and animals alike and, unless one can come to terms with this fact, it becomes increasingly difficult in later years to see the natural world with the same simple wonder of childhood. The pure, fragile water-lily, spreading its china coronet to drink in the sunshine of a summer morning, is born from and sustained, through its green umbilical cord, by the black ooze of the pond bottom. Yet even as one watches, it may be destroyed in a moment by a questing waterhen. But until that moment – like the waterhen – it is an infinite wonder and a delight to the eye lucky enough to see it.

The cottage 1967

A Hole in the Ground

My wife and I leaned on the sill of the little dormer window in the cottage roof and looked out westward across the Hampshire valley. It was October and still golden and warm after a hot summer. The lawn below was fairly flat for about thirty yards and then it sloped gently down to the valley floor where yellow reeds rustled in a plantation of tall poplars that stretched across the prospect from side to side. On the left was a broad herbaceous border and a cupressus hedge which divided the garden from a little field called

Church Paddock. The land followed the same gentle contour as the lawn and beyond that was a farm road and a tiny village church crouching over its precious chained bible among dark yews. To the right another small field extended as far as the end of the plantation before giving way to miles of unspoiled country to the north.

Before us the Test Valley stretched away perhaps a mile in width before rising again up gently sloping fields to the wooded horizon, where the sunlit roofs of a few farm houses amongst the trees were the only signs of habitation. The river itself flowed down the centre of the valley, winding in great crystal-clear loops south to Romsey town, past the great abbey church and on down to Southampton Water. Between the river and the cottage a carrier (the name given to the man-made side streams of the River Test which are famous for providing some of the best dry fly fishing in the world) followed the same southerly course.

The years had drifted by since I threw my last sandwich crust to the ducks at Raby Mere. A war and late student days, six years teaching at a College of Art in the Midlands and the concentrated effort to establish myself in the precarious world of the freelance artist had left little time to linger at the water-side. But I had been fortunate and freedom is its own reward. I had found myself able to live wherever I chose and had moved with my wife Rhona and our two children to the lush Hampshire countryside in 1959.

Eight more years had passed since then, and I had been very lucky a few weeks earlier to buy a piece of this lovely river and the old cottage overlooking it.

But I had been surprised and disappointed to find that the water was not visible from the window or even from the garden. There was nothing I could do to move the cottage nearer to the water but there seemed no reason why I should not bring the water closer to the house. I realised suddenly that this was the opportunity I had waited for for so long: the chance to make my own little lake and recapture the faraway dreams of childhood. Water was available and could be brought across the fields into the garden. It was going to take a long time and a great deal of hard work but, one way or another, I was determined to do it.

The cottage, too, was going to need a lot of attention, but the southern half of the building was in reasonable condition and the single storey section to the north could be dealt with when we moved in. I already had some experience of renovating and reconstructing old cottages, and even an old, dilapidated Cornish water-mill a few years previously, but this was the first chance I had ever had of working on a pool of any great size. I had decided to get the first part dug before we moved into the house.

We were not in the least put out, therefore, as we stood in the empty bedroom and looked at the view that October day, to hear the roar of a heavy engine and see the ungainly jib of a dragline swing its great toothed bucket into a small patch of water at the lower end of the lawn. The plume of black ooze that rose high in the air and fell on the green grass like an ink blot on a clean sheet of paper made me thrill with pleasure, for all beautiful things spring from primeval slime and I was watching the birth of the lake I had dreamed about so long.

Anyone who has ever dug a large hole in the ground will know that the soil which is produced seems to have twice the volume of the hole itself, and this was so here. After a few days the piles of earth became so great that they had to be moved by tractor and trailer and spread over adjacent fields. The clearing continued at the same time as the digging and in about three weeks the first stage of the pool was complete. It was roughly pear-shaped.

There was at this stage only the one small island and my main purpose in leaving this was to provide a safe place for ducks to nest. I wanted to produce a pool as nearly natural as was consistent with keeping it accessible and reasonably manageable, but I was soon to learn that even relatively tame birds do not co-operate with human plans unless they are restricted by cages, high wire fences or other unnatural devices.

As the work proceeded the whole area became a confusion of piled-up peat and black water, but it was unbelievably exciting at every visit to see the water area getting bigger. The plantation of poplars had stretched right across the western side of the house and, in order to open up the view and give more space for the lake, it

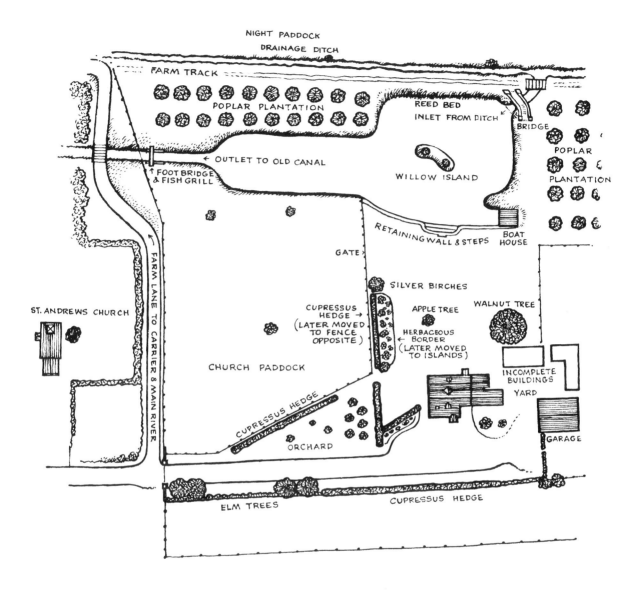

NIGHT PADDOCK

DRAINAGE DITCH

FARM TRACK

POPLAR PLANTATION

REED BED

INLET FROM DITCH

BRIDGE

POPLAR

PLANTATION

← OUTLET TO OLD CANAL

↑ FOOT BRIDGE
& FISH GRILL

WILLOW ISLAND

RETAINING WALL & STEPS

BOAT
HOUSE

GATE ×

SILVER BIRCHES

WALNUT TREE

CUPRESSUS
HEDGE →
(LATER MOVED
TO FENCE
OPPOSITE)

APPLE TREE

HERBACEOUS
← BORDER
(LATER MOVED
TO ISLANDS)

INCOMPLETE
BUILDINGS

ST. ANDREWS CHURCH

FARM LANE TO CARRIER & MAIN RIVER

CHURCH PADDOCK

YARD

CUPRESSUS HEDGE

GARAGE

ORCHARD

ELM TREES

CUPRESSUS HEDGE

was necessary to fell a number of these half grown trees. The difference in outlook that resulted from the operation was impressive. It was like drawing aside a curtain.

Whilst the machine was working the water remained as black as ink from the peaty nature of the soil, but within a day or two it began to clear and at the end of a week it was as transparent as crystal.

The water table of the valley is only about two feet below the general level of the soil in this area and the pool existed as soon as digging began. There was no need to bring in water to produce a lake. It was obvious, however, that if fresh water could be introduced through a stream the pool would be better for the movement and oxygen supply which it would bring. A channel was dug at the north-west corner, therefore, to connect with an existing ditch which was part of the drainage system of the area and another channel was opened up to the south which connected with the old disused Southampton to Andover canal. Only a small amount of water flowed through at this stage but it gave the pool life and movement at each end and prevented the water from stagnating.

There are plenty of signs of the old canal, including a well-preserved bridge in my neighbour's garden, and much of the waterway still runs south to Romsey. To the north the depression is visible in the fields in several places but within my own garden it is difficult, even with the help of old maps, to decide upon the exact line of this once controversial and disastrous project. One thing is certain, between 1794 and 1800 the canal ran across my garden either through the area of the lake or along the line of the stream which I have since opened up, and I find it fascinating to imagine the boats and cargoes and the canal people who passed along this way so many years ago. When digging, we came across some large pieces of almost petrified timber which might well have been part of the canal at one time, but it has not been possible to identify anything for certain.

When the digging of the first section of the pool was complete, we started work on the cottage. We were living at Braishfield village, about two miles away, at the time and I visited the place almost daily. There were so many things to do that it was often difficult to keep to an orderly plan but it was always interesting and exciting. On Christmas Day of 1967 my wife and I could not rest after our dinner and we went along to the cottage to walk about and digest our heavy meal. In a very short time my wife was hacking away at a hawthorn hedge which was about eight feet high and I had started carrying stones from the yard down to the water's edge. I planned to build a retaining wall at the bottom of the sloping lawn and some steps down to water level, but I had not intended to start the job on Christmas Day. I achieved only a token pile of stones, for they were very heavy and I had not brought a wheelbarrow with us. Nevertheless I had made a start and that was important: I find that the most difficult part of many jobs is beginning them.

Token or not, the stones must have weighed nearly half a ton and the hedge trimmings were piled about four feet high by late

afternoon. So we returned to the tinsel and holly at our Braishfield home with a feeling that we had probably done more outdoor work than most people on that particular day.

I had arranged for several lorry loads of random stone to be delivered to the yard. Demolition companies and builders are often contracted to dispose of stone and other materials when buildings are knocked down and it is usually possible to arrange with them to deliver such material to a new site rather than to dispose of it on an official dump. The cost of buying stone in this way is very much reduced and I had a good supply heaped in the yard where the lorries had tipped it. It was the rubble of an old pub, part of which had been demolished. The pieces of Portland stone varied in size from partly dressed pieces weighing more than half a hundred-weight to small, irregular-shaped lumps of only a few pounds. It was rough in texture and, having been weathered for many years, was attractively varied in colour by the presence of mosses and general exposure to the elements.

The plan for the wall was firmly fixed in my mind. For several years I had been meeting some of my friends from Braishfield village at the pub on Friday evenings and Bill, Douglas and Bob had always been regulars. From time to time we would vary our discussions on politics and religion and how the world should be run by debating more personal (and usually more interesting) problems. Whenever I had a new project in mind I would put it to them over the frothy amber glasses and, after we had searched our pockets for a pencil, we would draw out our plans round the advertisements on the beer mats.

The plan of the pool had been thrashed out this way and we had debated the retaining wall for some weeks during the autumn meetings. The consensus of opinion was that if the foundation stones were heavy enough I could dispose of the appalling difficulty of laying concrete below the water-level. I was easily swayed in favour of this theory because the high water table made it impossible, without pumping equipment, to lower the water-level in the newly dug pool.

Soon after Christmas I started building the wall. Bob, as with so much of the later work on the pond, helped me enormously with his labour and often with equipment like the sack truck which made it much easier to move the heavy stones; but most of all with his companionship, for I have noticed so often that two people who enjoy each other's company can do at least three times the work of one man labouring alone.

The edge of the lawn where it met the water was shaped and trimmed by hand with a garden spade and the foundation formed from the large more-or-less rectangular stones which were lowered into place with ropes and rough levers, and by any other method which would control them enough to prevent them falling away into the water. There was not much width of solid ground on which to stand between the slope of the bank and the water, for the wall was to be only about one and a half feet thick and we had many alarming moments when the great stones threatened to carry us with them into the water.

Stability was achieved by digging a place for each stone, so that it leaned backwards against the slope and we packed gravel and earth tightly between and behind the blocks. Once this line of stones was firmly in place the task became easier for we were then working above the water line and had a solid and more-or-less dry path on which to build.

Planks were laid down the slope of the lawn to prevent damage and provide a smooth path for the barrow loads of stone. This worked so well that we had to set up a buffer to stop barrows overshooting into the water and taking us in with them. Rather surprisingly, this buffer also seemed to do what we expected of it and only a few times did we hit it too hard and shoot a great block of stone into the depths.

When the walling and steps were complete, they formed a shallow curve which bracketed the edge of Church Paddock and the small inlet at the north-east corner where I planned to build a boathouse.

It is always a surprise to me, when looking later at work I have done with great expenditure of concentration or physical labour, to see how small and insignificant the results of it all are. When the stones were finished off with a cement binding between the top pieces the wall looked rather neat and tidy, but no sense remained of the bulk and weight of the individual blocks which had caused so much heaving and puffing. It could only be seen properly from the other side of the pool and it looked orderly: a long low wall built of little sections of stone, apparently no more difficult to put together than a jigsaw puzzle.

"Outlet grill"

The Splash of Clear Water

We moved into the cottage in March. Now we looked out upon water sparkling in the early spring sunshine. I was anxious to put trout in the pool as soon as possible and, in order to keep them from escaping into the surrounding ditches and waterways, it was necessary to fix a wire mesh at the inlet and outflow points.

Waterhens and a coot were swimming about on the new stretch of water within a week and I was shaken one morning to see a jack pike lurking below the newly built wall. Not being interested in the prospect of supplying fresh trout for the benefit of predatory pike, I determined to seal off the pool as quickly as possible. I now know that gratings are useful to keep trout in a certain area once they are of reasonable size but they do not prevent pike from getting in. It is true, of course, that only tiny pike can pass through, but they do this in alarmingly large numbers in this area and, once inside, they are like wolves in the sheepfold and grow quickly on the rich feeding.

The weather was remarkably kind to us that early spring and Bob again helped me. There was already a bridge built of railway sleepers across the ditch just before it turned into my pool and it was only necessary here to cut a piece of strong galvanised mesh to size and slide it down against the bridge piles to seal off this end from big pike invasions and the exodus of trout.

At the outlet end, however, we had to reinforce the banks with heavy wood and put a sleeper bridge across to form a firm structure on which to set our net. A few days after this defensive grill had been put in place I was working on a handrail along the bridge when I heard a twanging noise which I could not immediately locate. I looked down at the wire mesh, the vertical wires of which were about one inch apart, and was astonished to see a shoal of small fish charging the wires in considerable numbers. They were moving at such a pace that they were able to force their way between the wires and flick on into the deep water like a piece of soap might shoot out of a grasping hand in the bath. They were doing this so successfully that the wires were making the odd twanging noise I had heard.

There was a ripple on the water surface and the fish were moving so that I could not identify them. I decided later that they must have been roach because before the end of that summer I saw shoals of them from time to time, basking on the surface on hot sunny days.

[27]

My daughter Penny had a pony called Tonda. He had now been brought to his new home and was installed in Church Paddock. Scaffolding had been erected round the cottage so that work could be done on the roof. Workmen were busy removing tiles and I was still doing some joinery on the outlet bridge when suddenly there was a tremendous splash.

I looked up in time to see the pony's head surfacing in a welter of spray. He blew a mist of water from his distended nostrils like Moby Dick and thrashed about as if he were drowning, then he struck out for the bank. I could not imagine how he came to be in that situation and by the look of the builders who were all watching in astonishment, neither could they.

He managed to get his knees up on to the bank but was obviously in difficulties. I saw my son, David, up near the cottage and, shouting to him to fetch a rope quickly, I jumped off the footbridge and ran round towards the struggling animal. As a matter of fact, the only rope we possessed was the clothes-line and that could hardly hold a pair of wet socks on a windy day. When I got to Tonda he was lashing out madly with his hind legs, trying to get some impulsion to push himself onto the bank; he looked like a big fat schoolboy in the swimming baths practising leg movements while holding onto the side.

David had disappeared on the fruitless errand of trying to find a non-existent rope and the workmen were finding more comfortable perches from which to watch the spectacle. I couldn't hear what they were shouting but I could hear their laughter well enough. As I arrived at the water's edge the pony gave me a hopeless sort of look and slid backwards into deep water.

He shot back to the surface like a performing dolphin, enveloping me in a wave of water. Peals of laughter floated down from the roof as I grabbed his forelock and held grimly on. His legs were going like a stern wheel paddle steamer and there was a desperate look in his eyes as he slid slowly backwards once more, pulling me with him. I let go just in time as he went in for the third time and I heard tiles falling from the roof as the next wave of water hit me.

This time he turned away into the middle and swam easily up the length of the pool then, turning by the island, he struck out towards the newly built steps and trotted up them with a curtain of water pouring from him. Thunderous applause greeted him from the hysterical builders who had now descended to the ground to fall about in safety. Tonda kicked up his heels and galloped round the lawn in a mist of flying droplets, then settled to cropping the grass as if nothing had happened. He dried out more quickly than I did, but after returning him to his own field I was clapped back into the house by an appreciative audience of happy artisans.

In April of that year I put three hundred trout in the pool and they brought the place to life immediately. Young trout rise much more readily than mature fish and leap about like children in a school playground. In the summer evenings, when the sun was low over the rim of the valley and the plantation trees cast long shadows across the lawn, the pool would glitter with fish leaping into the sunshine from the dark water.

I bought a sack of floating trout pellets and fed them daily. In a very short time they learned where the food supply came from and would swarm around the steps as soon as they felt the vibration of footsteps on the lawn. Whenever I went near to the pool, whether at feeding time or not, I could bring the ravenous shoals to the bank merely by stamping my foot.

Rainbow Trout

This first batch of fish was made up entirely of rainbow trout, although native brown trout were introduced later. Rainbows grow more quickly than brownies but they do not live as long and

Brown Trout

although their name (derived from the iridescent oil-on-water colour along their flanks) is rather more exotic than that of our native fish, I think the brown trout is a far more beautiful creature.

Its colour varies a good deal but it is much more attractive than the general silver grey of the rainbow. When taken from the water, a brown trout is more beautiful than anything any jeweller ever created. Its dark back shades through silver to rich olive green and ochre and its spots are larger than on the rainbow and brilliant red among the black. Laid on a river bank it is breathtakingly beautiful and I find one must need it rather urgently for the table before it is possible to kill it.

Man himself is as much a part of the order of things and as natural a predator as the lurking pike or the hunting weasel. He has hunted and preyed on other species since he inhabited this incredible planet and yet . . . There is for me always that sense of uncertainty, that moment of doubt when I take a fish from its natural element.

The whole question of killing is very complex and disturbing. Although life for all creatures in the water is precarious in the extreme, an endless sequence of kill or be killed, it is sobering to see the glazing of the bright eye and the fading of the brilliant colour when a fish dies. It is a bright existence gone forever from the water.

Angling is a highly popular sport and most lakes and rivers would quickly be stripped of their fish population if action was not taken. All responsible fishermen, therefore, take great care to see that the fish they take from the water are replaced by fresh stock. On the whole this is not a purely selfish attitude either, for the more one fishes the more one comes to admire and respect the fish and to feel an obligation to preserve them for future generations to enjoy.

Once the trout were firmly established in the pool I decided to introduce ducks as early as possible. I bought a little book on the subject and discovered the existence of call ducks. These pretty little birds are almost exactly the size and shape of wild mallard but are white, with ochre-yellow beaks and legs. Their main purpose it seemed, from what I read, was to act as decoys to bring wild duck down on to a lake. This they do by their habit of quacking loudly, particularly at night. They sounded exactly the sort of ducks I needed to start with.

Angela, one of Penny's school friends, was staying with us for the weekend and we all drove into the New Forest to a game farm which advertised in the local paper. There were many hundreds of ducks available in large out-door pens and we bought four pairs which I thought would be enough to start a breeding group of our own. One wing of each bird had been pinioned by the trimming of the flight feathers. This seemed the most sensible way to restrict the birds' freedom to fly away because it is a temporary precaution and as the flight feathers grow again after the moult in summer there is no permanent mutilation of the birds.

The feathers of these little ducks are as white and smooth as newly fallen snow and in bright light they reflect local colours of pale yellow and blue and green, according to the light and where the bird may be.

We put them into cardboard boxes on the back window shelf of the car, piled in and started for home. It was a warmish day and, before we had gone more than a few miles, an ominous silence fell over the passengers. Ducks, in any confined space, have a powerful smell which none of us had ever suspected before. Another mile and I was forced to pull off the road and stop the car. The doors flew open and everyone fell out gasping. We made the rest of our way home with all the windows down and it was just bearable.

It was an exciting but nervous moment for all the family when we released the birds by the water side. We were taking a big chance, for I had no idea whether they would stay or swim away down the ditches and side streams, never to return. When the lids were lifted they sat still for a moment in the sudden brilliance of light then leapt flapping and quacking out of their containers, across the grass and into water. It was the first time in their lives that they had been free of a wire enclosure and the first time they had ever seen enough water to swim in.

There had been reeds growing on this ground before the pool was dug and they were already tall and thick again on the far edge of the pool. It was into the shelter of these reeds that the ducks bolted and hid themselves away. The show was over and we left them there to decide for themselves what they would do next.

We all left the garden in order not to panic the nervous birds but we watched carefully from the cottage windows. It was at least half an hour before they ventured forth one at a time and began to explore the edge of the reed bed. At last, when they were all in view they were seized with a sudden ecstasy of joy. They chased around, flapping madly in showers of spray. They jumped in the air and dived below the surface of the water and the loud quacking sounded like the crazy laughter of long-term prisoners who have found their cell doors open and the wide world stretching before them.

From that moment on the ducks were established and although, since the first summer season, they have all been allowed to fly freely, they have never shown any sign of deserting the pool permanently. This policy of leaving all the ducks unpinioned has meant that they spread far and wide, particularly at nesting time, and there have been heavy casualties over the years, but even those that depart during the day fly back in as the sun goes down, and it is a joy to see them arrive and depart like planes at a busy airport. Once the call ducks were firmly at home I introduced several pairs of mallard in the same way and, perhaps reassured by the confidence of the resident white ducks, they too settled in quickly and have stayed with us ever since.

Even before the introduction of the ducks from the game farm we had been delighted to see several plump little black and white tufted ducks on the water on several occasions and we were astonished that

very first spring to find that a brown female had nested on the bare island behind a clump of kingcups which I had planted there. It is common now to have a raft of these pretty creatures every winter. They have never become tame as mallard will easily do but, being diving ducks, they obviously find the deep water to their taste and they stay for months at a time, often twenty or thirty of them with a few pochard here and there among them. One day this winter we counted forty-five of them bobbing up and down on the choppy surface of the lake.

Just as our own ducks are free to come and go as they please, we often have very welcome strangers who visit us for short periods and then go on their way. We have provided temporary accommodation for pintail, teal, wigeon, Australian wood-duck and mandarin duck at various times and swans and Canada geese are regular visitors.

Piper's Moon   One of the delights of living by water in a relatively undisturbed
area is the great joy of lying in bed and listening to the night sounds
from the pool. In the velvet quiet the smallest fall of water over
stones can be heard like distant tinkling bell notes. The waterhens
stab the night with loud liquid croaks, repeated twice, and ducks
chorus suddenly together like carousing revellers laughing at a
risqué story. The high-pitched cry of a coot comes from the reed
beds or the warning metallic cry of alarm, but most delightful of all
is the cascade of rippling, watery notes from the little grebe. It is
like hearing the pipes of Pan, bubbling and clear, yet delicate and
vaguely melancholy, like the cry of the curlew on a lonely moor. It
is the sound of pure enchantment and when all is perfect the sedge
warbler will add his song to the bewitching night.

Strong wings can be heard too when the moon is bright on a
silver valley. Wild geese circle low round the house and wild duck
whistle in on fast pinions and plane down to set the water splashing
in the deep shadows.

Sometimes, when I have been working into the early hours of the morning, I feel restless and disinclined to go to bed. On these occasions I often switch off the lights and wait for a short time in the darkness of the room, then pull aside the curtains gently so that the sound cannot be heard outside. It is astonishing how light the night world seems after the total blackness within. I open the doors very quietly and step as silently as I can out into the cool air. The night is alive with a thousand muted sounds. Under the great spreading branches of the walnut tree, pressed against its deeply ridged trunk, I seem to be camouflaged from the little creatures who hide and sleep away the daylight hours in the tangled undergrowth of the plantation. They are abroad when humans are asleep and the world is theirs again as it has been since the dawn of life.

Rabbits nibble the lawn grass and hop about the hedges. They are busy making the dozens of little scrapes that are so irritating to those who care about the perfection of their gardens. Hedgehogs snuffle and grunt noisily as they rummage about for tasty morsels. A beetle drones past like a buzz-bomb on its way to fresh ponds and ditches and bats flit by on their tiny umbrella wings.

The world goes silent around me as a rumble like distant thunder drifts up the valley. It increases in volume until the rhythmic clicking of wheels can be heard. It is a train going north from Romsey on the far side of the river. Lights flicker through the trees and pass like a string of bright fireflies as the ground vibrates gently. They blink and disappear into the darkness as the beating of steel on steel dissolves into the night. Silence returns like a gentle wave. A rabbit hops up the little knoll by the silver birches and a coot calls in the reeds. The busy life and death of the wild creatures is in full swing again and the orderly bulk of the silent cottage seems strangely out of place in a primeval world.

Where birds and animals are concerned, there is always something new to learn. I quickly discovered that ducks, for example, are crazy about fish food. The fish were in the water before the ducks were on it and the daily routine of feeding them was well established before the call ducks arrived. I was delighted to find that the ducks regarded the water as their own but it had not occurred to me that they would regard anything thrown into the water as theirs too. They did, and the battles that followed were spectacular for the fish liked their food too and had no intention of having it stolen.

Trout pellets are nutritious and expensive – far too expensive to feed to ducks who are well able to fend for themselves. Besides which, they were given their own rations on the lower step by the water where they could scoop the grain without competition, for the fish could not get at it there and the bantams which we kept never seemed happy too close to the water. The ducks regarded this grain as an entrée, however; a mere appetiser for the main course of trout pellets which was to follow. They quickly became the most expensive ducks in the locality, for although the trout reacted to the competition by increasing their speed and acrobatic displays, they were much smaller and lighter than the ducks who, in any case, were willing to fight all comers for the magic pellets.

The trout grew at an astounding rate and the competition became fiercer month by month until the balance tipped in favour of the magnificent and muscular rainbows. I was never able to find a way of discouraging the ducks from trying to steal the fish food and I had to leave them to fight it out. A fish of several pounds, accelerating from deep water, is a fearsome missile as it breaks the surface and the ducks, crazy to grab the floating morsels, were frequently shot several feet into the air by the impact.

Many of the original fish fell victims to a wide variety of predators, as I had expected. They were taken by herons, pike and even marauding cormorants, apart from occasional anglers, so that the numbers had dropped considerably as far as I could tell. There were, however, still many times more fish than ducks – perhaps well over a hundred – by the time they had reached the heavyweight class and they seemed to grow faster as their numbers declined. As several fish reached five and six pounds, feeding time became spectacular in a way which I would have found difficult to believe had I not witnessed it.

On many occasions the ducks which were still willing to charge into the flying maelstrom escaped death by a miracle, for they were often grabbed by the beak and hauled away down into the dark water. Several times I had decided that a duck must have drowned before it finally popped back to the surface and dashed away to safety.

I cannot be sure, but I believe that these incidents were not deliberate. The pellets floated on the turbulent surface and could be clearly seen by the fish below; they could also be seen just as clearly by the ducks from above. In the turmoil it stands to reason that the chances of a fish and a duck going for the same pellet or group of pellets at the same moment must be high. A trout's mouth is much larger in proportion than a duck's bill and is full of minute needle-sharp teeth. Trout take pellets in a slashing arc of great speed and once their mouth has closed on food, or even if they have missed it, they must continue the downward curve. In these circumstances, once a trout's teeth have closed on a duck's bill, the consequences must be inevitable.

As the trout became bigger and less in number I was more and more reluctant to catch them for the table. They were so tame that they would cruise close in to the steps and seemed more like pets than wild fish. I knew several of them individually by scars on their backs and flanks and we gave some of them names.

Sometimes, for the interest and amusement of visitors, we would throw in several full slices of stale bread. The ducks didn't even attempt to compete on these occasions for the slices were seized immediately from below and the light patch could be seen speeding this way and that in the deep, dark water as the fish which grabbed it was chased by the others. After a few moments of this softening-up process the bread would break up and the bigger pieces would flash away on separate courses until it all disappeared into speeding fish. Three or four slices in succession would be taken in this manner and all would be gone inside thirty seconds.

Rainbow trout do not live as long as brown trout and even in the third year their condition seemed to deteriorate. There were signs of fungus on some fish and several had succumbed to the snail-borne disease which causes blindness.

I decided to net the pool out and release the bigger healthy fish into the main river. Some of the rainbows were now four years old and many of the big ones were still in splendid condition. It was an interesting operation and took four of us a whole afternoon to complete. We used a long rectangular net with a rope attached to the top at each end. Cork floats kept its upper edge on the water surface and lead weights ensured that it remained in contact with the bottom of the pool when it was drawn along. One end of the net was held by its rope close in to the southern end of the wall and it was stretched across to Willow Island where the other end was held by a man who had rowed across for the purpose. The sweep began by moving the net slowly in an anti-clockwise direction round the edge of the pool whilst the man on the island acted as a pivot in the centre.

Although it needed care to keep the net fully extended, it was a simple circular movement which continued unbroken until the moving end reached a point on the bank opposite to the end of the wall where it had begun.

From here on the pool narrowed and it was necessary for the man on the island to row back to the wall, hauling his end of the net with him. Some fish probably escaped during this manoeuvre but the net now stretched from bank to bank and the operation was completed by pulling it slowly along towards the far end of the pool.

When we were about fifty feet from the grating on the outlet sluice the water surface began to seethe. Trout leapt from the turmoil in panic and a few escaped over the floating edge of the net. The catch was very heavy. There was no question of hauling the big net on to the bank at this point. The fish were lifted from it in strong hand nets and sorted immediately into a series of water tanks which we had already placed at the spot. We worked for more than half an hour before the big net was light enough to be pulled in and the last of the fish disposed into the tanks. We counted more than a hundred and twenty rainbows, eight pike, six eels, about forty smaller brown trout, dozens of smallish roach and three magnificent, glittering golden orfe which my children had bought for my birthday several years before.

The orfe and smaller trout were returned to the pool and nearly a hundred big healthy rainbows were released into the river.

I heard many tales in the pub and on the river bank that season of unusually magnificent rainbows caught up and down the river. It made me a little sad for I was sure I knew some of those fish by name, but it was not all loss. I'm sure that most of those fish would enjoy a period of freedom in the river before they were taken by some predator or died and floated away with the great rafts of weed to the open sea.

The Boathouse   Swallows came to the boathouse the second spring after it was
completed. It was April and lambs were bleating and gambolling in
the home field over the cupressus hedge.

The field was surrounded by sheep wire, but from time to time
we would surprise a pretty little black-faced creature in the orchard
or in the flower border by the gate. It would panic immediately and
rush in bouncing leaps back towards the pasture. But although
these lambs had had to struggle through the six inch square mesh of
wire to get into the garden they seemed unable to see it on the way
back. They would dash headlong into the taut wire and be
catapulted back several feet, stagger back onto wobbly legs and
make another blind rush. The effect was both comical and rather
alarming but they never seemed to come to any harm. After three or
four dashes at the wire they always managed to find a hole with
their hard little heads and would catapult through with a velocity
that left the wires singing like plucked violin strings. An hour or so
later another little face would pop up, wide-eyed, from behind a
clump of daffodils and the wires would be set twanging again as
panic sent it streaking back to its anxious mother.

Wild violets peeped like purple confetti from beneath the hedge
by the cottage windows and sticky buds of the young horse chestnut
in Church Paddock were bursting into leaf. In the hedgerows the
blackthorn was white with blossom and Tonda, Penny's New
Forest pony, sniffed the spring air with flaring nostrils and
thundered around his field like a medieval charger sending up
showers of divots from the green earth. The world seemed full of
birds.

The wild ducks were nesting and I was watching the first string of ducklings to appear on the pool that year. They moved in line ahead like ten tiny camouflaged corvettes following a battle cruiser. I did not know where that particular duck had nested, for they spread out every spring after mating and find their own secret places along the streams and ditches and no amount of careful scheming on my part will persuade them to lay their eggs in safer places. We were all aware that the anxious teeming time of beautiful baby creatures was about to burst upon us once more.

The lake in spring is as exotic and savage as any tropical jungle. Every day seems to produce more and more newly-born birds and animals – each one an exquisite masterpiece of jewel-like design and each one quick with the spark of individual existence. It is a microcosm of the whole world, a cross-section of the precarious but persistent balance of natural forces. Every year the fact of new life springing only from death must be grasped anew in order to appreciate the beauty of it all. Otherwise the great wave of new creation meeting the inevitable counter-surge of savage destruction can be almost unbearable. Beneath the shining water surface the upsurge of life is even more prolific, the battle for survival even more intense.

Young fish of all kinds are present in their thousands and provide food for animals and birds. There are predatory insects that will attack and eat the small fry also. The larger fish feed upon insects and upon the smaller fish and all kinds of eggs. Pike will take almost anything – including voles and ducklings in addition to other fish.

A handful of weed taken from the water and shaken on the step will disgorge dozens of writhing insects – sometimes hundreds. There may be fresh-water shrimps and beetles, stick insects, scorpions, spiders, snails and nymphs, and the larvae of a wide range of aquatic flies.

There will be thousands of other creatures so small that they remain in the film of water on the weed surface; water fleas no bigger than a pin-head and others so tiny that they are invisible to the naked eye. But they are all links in the complex food chain and depend upon the weed and each other for their continued existence.

The cuckoo was sounding off across the valley when I caught the first flash of blue-black wings disappearing under the low entrance to the boathouse. The swallows were building. The first nest was tight up in the rafters under the old tiles, constructed like a sculptor's model with pellets of clay one on top of the other, pressed into place with care until the shape was just right. The parent birds brought up two broods that first year and the little bottle-green boat was spattered with white spots so that the seat had to be washed down each time it was used to visit the islands.

The miracle of migration is brought home to us in more senses than one with the delight of seeing our own swallows return each spring, or perhaps it is the next generation returning to the tiny cup in which they were born last year – who knows? Our boathouse is certainly now the northern terminal of a line which reaches south across the equator to the African heartlands or even the Cape and when the swallows return again it is a high point of the year. There are two nests side by side in the rafters and another on the wooden lintel over the door. We feel quite anxious each spring in case anything should happen to our swallows on the long journey home.

The boathouse looks old although it has been there only about eight years. Building it was quite an adventure. Its venerable appearance is partly due to pigs. As the pool and its various features grew, Bob was an almost continuous helper, adviser and, very often, source of necessary materials. He had a thriving little farm up on the top of Fern Hill. On a clear day it was possible to see the Isle of Wight from the highest field. His copses were alive with pheasant

that flew in for protection when the guns of the surrounding shoots crashed through the winter afternoons and in the spring the woods shone with acres of wild daffodils. Bob specialised in pigs.

He is a tireless worker and he built all the pig houses himself. In order to carry out this Herculean task he had assembled a vast quantity of materials which he kept under a series of low sheds. It was an Aladdin's cave of useful and often fascinating odds and ends and he was always ready to scramble around trying to find any special materials I might need at any time. We had many interesting sessions rooting in the dim recesses while the sun beat or the rain hammered down on the tin roofs. Those sheds had the fascination of the old type junk shops and, whatever the current need might be, they always seemed to contain the special piece of iron or timber necessary to complete the job.

Not only would Bob extract each particular item from the piles of railway sleepers, stacks of iron sheets, metal tubes, rolls of wire or random timber, but he would insist on carrying it to my car and fixing it on the roof rack. He would be genuinely hurt if payment was mentioned. His workshop/garage was also stacked with 'useful' oddments waiting for the day when they would come into their own. So crowded was it that his fan-tail pigeons had taken it over as a nesting colony. They built their tatty nests all over the place between piles of tools and mysterious collections of bars and wheels and implements. His car was forced out into the yard for most of the year. I remember that car very well. It was a dark green Ford Consul with a white stripe on the nearside front mudguard just behind the wing mirror.

Bob is not a car lover and his vehicle had a lived-in kind of look, but the white stripe was kept fresh by the pied wagtail that created it. He seemed to spend most of his waking hours there making aggressive and threatening gestures at his reflection in the wing mirror. He had either a particularly strong hang-up about territory or he was just a bit dim. His attacks were so constant and vigorous that we wondered why his face was not flat.

Whenever a new project was being considered I used to put the problem to Bob, Bill and Douglas in the pub on Friday evenings and we would hammer out a solution with a pencil on the beer mats or with a finger on the wet table top. When the boathouse project was introduced Bob sounded his usual warning, 'Don't go out buying timber – let's have a look in the sheds. I might be able to find you something.' It was not just that he saved me considerable time and expense – which he did – but once Bob was in on an idea it was like having a travelling companion in unknown country and one who, when you felt tired, would offer to carry you on his back.

Bob did find me something. A big stack of two-inch thick random timber which had once been part of the piggery. It is doubtful if any force of nature can take the newness off anything more efficiently than the jaws of a pig. The planks were perfect for the job – they already looked ancient, but they were strong and sound. He suggested that the boathouse should be built in his big piggery where we could work under cover and with electric light when necessary.

An area of flat concrete floor was cleared and the foundation laid with railway sleepers. We spent many happy evenings sawing and hammering and doing mathematical calculations with a lump of chalk on the floor or the side of the pig pens.

Much of the work was done on cold winter evenings, often with rain beating down on the roof, but it was cosy there in the long, low building with several hundred pigs keeping the temperature comfortable and the air vibrant with contented squeals and grunts. As the building took shape so did certain hazards which kept us alert. The structure was put together with large nails but as it all had to be taken apart again for transporting back to the site on the lake, the nails could not be driven home. They were knocked in just far enough to hold things loosely together but not so far that the timber would be difficult to separate. The result of this was that the whole thing bristled with nails like a geometrical hedgehog. After a few painful contacts we went about our work rather carefully.

When the rafters were put in place from the wall plates to the ridge plank it all began at last to look something like a boathouse and our increased confidence made us careless. By this time the shaky structure must have weighed a ton and the rafters alone some hundredweights. We stood back once more, lit our pipes and surveyed the results. It looked good. Old, but good. As we strode confidently back into the boathouse through the doorway, one of us bumped against the side and moved the building enough to extract all the temporary nails from the ridge plank.

We both saw the dangers in time but in our efforts to get out before the falling timbers hit us we jammed in the doorway and shook the pile so violently that the nails were pulled away all over the structure. It was like being an eye witness to the San Francisco earthquake.

Wood that could survive the attention of pigs was not likely to suffer much even from that nerve-shattering collapse, however, and when the dust cleared and the frightened pigs quietened down it was simply a matter of thanking God for our deliverance and putting the jigsaw puzzle back together again. All the pieces of timber were then lettered and numbered with a brush and a bottle

marsh marigolds

of indian ink ready to be taken down and made ready for transport. The whole thing was moved, about five or six pieces at a time, on the roof rack of my Hillman Imp until it lay in an organised confusion by the water's edge.

I had already had a small inlet cut in the north-east corner when the pool was first dug and once the foundation of stone and railway sleepers was in place it was a pleasure to reassemble the building on the site. With a chart showing the position of each numbered piece it was a simple task and the whole of the woodwork was rebuilt in a couple of days. This time the nails were driven in far enough to avoid any spectacular disasters, although Bob insisted that each nail should be left about half an inch proud of the woodwork until the whole thing was complete – 'just in case'. This will be recognised as a wise precaution by anyone who has tried to extract six inch nails from heavy timber.

Although I had played about with building various kinds of walls – particularly dry stone walls – for some years, I had never had the opportunity of tiling a roof myself. When the woodwork was complete and the rafters firmly in place it looked like a rather cute doll's house waiting to have the finishing touches added.

I bought about three hundred and twenty feet of one-inch batons (the only material I had to buy) and, after carefully working out the gap necessary to hang the tiles with two-thirds overlap, I fixed them horizontally across the rafters with galvanised nails. I was surprised at how rigid and strong these relatively flimsy strips of wood made the whole roof. It was quite strong enough to climb about on.

Some years before, when restoring a pair of old cottages at Braishfield, I had bought twenty-two thousand hand-made clay tiles from a demolition company in Dorset. (I still have two of them with the date '1788' cut with the point of a trowel in the wet clay. The tile makers' names are there too – James Brown and William Fance and what appears to be the wages list for the week; all done with a copperplate flourish and repeated several times as if to establish their immortality for all time.) About twenty thousand of these tiles were used to re-roof the cottages but I still had the remainder, which I had brought with me to Herons Mead. They were exactly what I needed for the boathouse.

Many of the tiles still retained one or two of the wooden pegs which must have been put in place a year before the French Revolution. Sitting on the little building there by the waterside, I found it pleasant to contemplate the events these pieces of clay must have witnessed over their long life. They are very beautiful. Each tile is curved and has an attractive rough texture. The colour varies from bright vermilion to dull Venetian red. They have the patina of almost two centuries of English sunshine and rain and are patterned with mosses in a wide range of emerald, apple and viridian greens. Any one of them, tastefully framed and hung in a London art gallery, would get rave notices from the critics.

In order to make each of the gaps between tiles match the centre of the tile below it was necessary to start each alternate row with a half tile and this was a tricky business. Even roof tiles, it seems, must pay a price for beauty and they are rather fragile. To cut them to shape without breaking them they must be chewed away piece by piece with a pair of sharp pincers, like nibbling a biscuit. But they are thick, tough biscuits and the job is tedious and very hard on the hands. Too big a bite and the tile will break right across.

Tiling the boathouse roof was a delight. The job was not too big, it was easy to get at and the surroundings are so beautiful. The missing pegs were replaced by short galvanised nails and as each tile was set in place the colourful, slightly irregular pattern grew like magic. It had the thrill of doing one's first weaving patterns with strips of coloured paper in the infants' class at school, and the fascination of doing work which I had so often watched skilled men doing before was as satisfying as being allowed to sit on a three-legged stool when I was a child and milk my first cow.

One of the nice things about building 'old' structures is that one does not have to be too mathematically accurate and small irregularities often add to the visual pleasure. By the time I arrived at the apex of the roof the line was just uneven enough to make the curved ridging tiles look interesting.

Occasional rises in the water level during heavy rains have caused the building to sink slightly and twist just enough to add to its charm. It doesn't seem to have given the swallows any cause for concern, nor the wrens for that matter, for we have caught them often in the cold days of midwinter, huddled together, six or eight in each nest, keeping each other warm.

They have always been inoffensive little squatters in the past – vacating the nests before the swallows returned from their winter holidays in the African sunshine. This year, however, they sealed up the open cup of one nest, leaving only a small hole through which they themselves could enter but which made it inaccessible to the owners. The swallows built another nest about twelve inches away and succeeded in producing three broods of youngsters there before they flew south once more. The wrens brought up two families of their own in the newly converted residence and I saw no sign of resentment on either side.

When the swallows come with the summer anyone entering the boathouse can witness a remarkable piece of aerobatics. The floor area is only eight feet six inches wide by ten feet six inches long and the foraging birds do not seem to notice intruders on their territory until they sweep down with beaks full of assorted insects, at what must be about thirty miles per hour, as they swing under the boat entrance. Confrontation is point blank and reaction is incredibly fast. There must be about twelve inches of air space in which to brake, turn and fly out again but it is done so fast that the human eye can hardly follow the manoeuvre. 'Bird brain' is often used as a term of contempt but the brain of a swallow must be a fantastic miniature computer to organise such complicated muscular reactions. It is an incredibly beautiful computer too, almost weightless, and whoever programmes it must be some sort of God. [53]

They never drop a single insect, as far as I can see, and if I retreat immediately out of the door they are back in a second, stuffing the gaping mouths as if nothing unusual had happened.

A waterhen nests on the wooden staging just inside the boat entrance most years but she is a different kettle of fish and waterweeds. She panics like all her kind, flings herself, splashing and flapping, through the opening and runs along the surface of the water to gain airspeed, leaving a line of bright rings decreasing in size from her first splash to the point where she is airborne. She swings away, low over the surface, to land on an island with a flash of white rump or splash clumsily down among a group of startled mallard drakes sulking in the reeds because there isn't a duck to be found.

The wooden staging covers about half of the water area within the boathouse, making it easy to step dry-shod into the boat, and it is a convenient platform from which to study underwater life. When the boat is out it has the great advantage that the water is well lit from the front opening but the roof cuts out the bright reflection of the sky and leaves a crystal clear section of water which can be studied as intimately as an aquarium. If one sits or lies very still even the fish do not seem to notice the human outline against the dark roof and will drift slowly in, only a foot or two from the eye, and explore the underwater foundations and muddy bottom for tiny snails and insect larvae.

Because the strong, direct light of the sun is excluded the waterweed is thin and sparse and presents an uncluttered stage on which to watch the players.

Pondskaters and whirligig beetles slide and spin on the surface and shoals of small fry drift into the comparative safety of the shallow water. The shoals move in unison, as if by some outside force. They form and reform in a kaleidoscope of patterns like iron filings on a sheet of paper animated by a moving magnet below, or like flocks of starlings coming in to roost in winter woodlands. The same strange mass movement is noticeable in the flocks of lapwings over the cold ploughland in the short, dark days of January and with the waders over the mudflats of the estuary when the flood tide creeps back to reclaim their feeding grounds.

Caddis fly larvae are also down there in their home-made tubular shelters and the fierce predatory nymphs of the dragonflies. Water-boatmen and beetles scour the weeds and pebbles for unsuspecting victims. It is a few cubic feet of teeming life and sudden death and a part of the endless natural cycle that turns relentlessly on. It is happening right now inside the little building that started life in the piggery at Fern Hill.

I doubt whether the old tiles have witnessed quite the same scene before in two hundred years and I cannot but hope that, when the boathouse finally collapses, someone will rescue the tiles to roof some other building and delight the eye of future generations for another hundred years.

Messing About in Boats

While the boathouse was being built I began to think about a boat. An advertisement in the local paper sent me off to Southampton to look at a likely prospect. It is surprising that past experience did not make me look more carefully but, for some strange reason, I bought it without close examination.

It was a small clinker-built wooden affair painted bottle green and blue. It was far from being new but I felt its chunky ruggedness would suit the well-worn appearance of the boathouse. It certainly did.

Two men delivered it about a week later and I was a little surprised at the fact that they were sweating profusely as they carried it down the lawn to the water. I was leaning nonchalantly on an oar, waiting for the moment when I could leap into my own craft and scull expertly round the island. They laid it down at the water's edge, mopped their brows and wheezed. Then, after flexing their shoulders they lifted it about six inches off the ground and dropped it on to the water.

It sank like a stone. It disappeared so fast that even they looked surprised. The wooden seat floated away towards the island. After glancing swiftly at each other the larger of the two men turned towards me, closed his eyes and held up two massive hands, palms towards me.

'Don't worry, sir,' he said. 'It's just that the wood needs to expand.' They started back to their lorry at a fairly brisk pace without further comment. I ran after them.

'How am I going to get it out?' I asked.

'Oh, don't try to get it out, sir,' he said, still walking fast. 'It's just that it's dried out a bit too much.' He got into the cab and started the engine. He leaned out of the window as the lorry moved away. 'Leave it a couple of days to expand and it will float a treat,' he shouted, as they rattled away in a cloud of dust.

I went back to the water's edge and looked down. The mud was beginning to settle and I could see the top edges of the boat dimly. The metal rowlocks glinted about six inches below the surface.

Two days later a thin film of scum had started to form and hundreds of tiny snails were already crusting the woodwork as I tried to raise the dinghy. After struggling for some time I understood why the delivery men had been sweating and I contemplated painting 'Wreck' on an empty oil drum, mooring it over the hulk and abandoning the whole thing. This proved

Rose Island

impractical, however, for the boat was lying where the boathouse was to be erected and would have been a permanent hazard to shipping.

With the help of my son I managed eventually to slide the bow on to the lower step, empty the water out and clean it up. The lorry driver was half right in his parting shot: the planks had closed up and when re-launched it did float, but it was never quite a treat. There was always a fair amount of dirty water in the scuppers and if it were left for any length of time even inside the boathouse out of the rain there was a lengthy bailing job to perform before it was fit to enter without Wellington boots.

After struggling with it for about three or four years I eventually abandoned it on Rose Island, which had been formed by that time, and it lies there still – chunky, rugged and a picturesque warning to anyone who might be foolish enough to buy a boat in a poke. It was entirely my own fault. I could not even excuse myself on the grounds of ignorance for I was not without experience of being messed about by boats.

Some years before when I first came to live in Hampshire and had found a common interest in fishing with Bob and Douglas, I was looking at a little lake north of Romsey. It was a warm sunny day and the water lay sparkling through the dark surrounding trees. The stillness of the afternoon was broken by footsteps coming down the lane and in a few moments I found myself passing the time of day with a man I had not met before. The conversation turned to fishing and I enquired whether he knew the owner of the pool. He asked why I was interested and I told him that I would like to know if there was any chance of renting the water. He told me that it was his lake and said that if I cared to stock the water with trout I was free to fish for them. This was to be my first introduction to the pleasures and problems of fishery management.

Bob and Douglas joined me in the scheme and in a few weeks we had put five hundred small rainbow trout in the water and there seemed nothing left on earth to wish for. As it turned out, there were a number of highly technical problems lying in wait for us. These were concerned mainly with lack of oxygen caused by deep mud, unsuitable weeds, lack of water supply and the effects of dense surrounding trees. However, although the episode lasted only a few months, the pleasures of that spring and summer remain to this day an endless source of recalled incidents and rolling laughter on winter evenings at the pub.

There was a boathouse in the corner of the lake and inside it, beneath the long snagging arms of encroaching brambles, was a boat. It was sunk. It lay there on the mud full of the rotting debris of more than one autumn and winter. Only the top edges were above water and little clumps of grass grew from joints in the woodwork widened by the effects of water and neglect. It was a pity that the boat was in such bad shape because on water surrounded by an impenetrable mass of trees and undergrowth there was no chance whatsoever of fly fishing from the bank. It was a simple enough problem, however. It must be taken from the water and dried out, cleaned and tidied up and the bottom would need to be tarred. There were three of us to do the job and plenty of young trout ringing the surface to urge us on.

On the subject of getting a heavy water-logged boat out of the water and upside down on the concrete floor in the restricted space of a small boathouse I will say little. At the time we could have said even less, we were so breathless and exhausted. After about two hours of groans and curses in a welter of flying water and stinking black mud it was out and so, I felt sure, was my spinal column. Bob had hurt his leg and Douglas was muttering about golf being a good way to spend one's leisure. We leaned against the boathouse wall

watching the trout jumping for about half an hour and went home.

About a week later the boat seemed reasonably dry. I went to a ship's chandler in Southampton and bought a gallon can of pitch. Apart from almost severing my fingers with the wire handle I got it safely home without incident.

When I prised the top off the can I was surprised to find, instead of a soft workable mastic, a hard glossy mass of what appeared to be black glass. Apart from spoiling the virgin brilliance of the surface with finger prints, my hands made no impression on the stuff. I took a six inch nail, placed the point on the surface and struck it with a hammer. It shattered into a million tiny slivers and left an ugly hole in the jet black lump. I removed several pieces from my hair and one from my left ear.

We held a conference on Friday night at the White Horse. Bill can always be relied upon to come up with the answer to any problem whether it concerns politics, religion or how to tar the bottom of a boat. 'It must be put on hot,' he said, and he was obviously right. Careful plans were laid for the weekend.

We acquired a metal oil drum and bored holes in the sides and bottom. Coal and wood were carried to the site and the boat was manoeuvred outside and laid bottom up on the ground. Bob had brought a big brush with a long handle. Douglas supplied the metal shaft of an old golf club. Bill, always the wise one, stayed at home.

I have been nervous of fierce conflagrations ever since my wife, Rhona, left a pan of cooking fat on the stove in the kitchen and came into the sitting room for a chat with weekend guests (they escaped unscathed but have not been back since), and I suggested that we place the brazier in a small clearing about thirty feet away from the building in case anything went wrong. It did not occur to me at the time that it was also some distance from the boat.

The oil drum was stuffed with paper, wood and coal and extra supplies were laid neatly to hand. The lid was removed from the tin of pitch and the golf club shaft pushed through the wire handle. This was then laid across the top of the oil drum with the tin dangling over the centre. Bob struck a match and pushed it through one of the holes. In no time the fire was roaring nicely and the pitch, after sending up a few puffs of blue smoke, began to melt and bubble. We stood round it with smiling anticipation like gypsies round a cauldron of stew.

With general approval Douglas dipped the brush gingerly into the bubbling mass and strolled across to the boat to lay on the first waterproof coat. He drew the brush back and forth along the bottom. Apart from a scraping noise nothing happened. He banged the brush to get it started but he might as well have used a hammer. The brush was a solid lump of shining hard pitch. Action was needed and quickly, because the fire was getting hotter and the pitch was boiling rather violently.

I was a bit younger than Douglas and Bob and somehow I felt it was up to me where quick running was concerned. I grabbed the brush from Douglas and dipped it again into the bubbling, hissing mass. I had to leave it in long enough to melt the brush. I ran like a hare to the boat and made a mark about four inches long before things began to stiffen. I pulled the brush away and long strings of black toffee rose from the spot and hardened instantly like the hair on a man who has just had a nasty shock.

In the next few minutes I made about a dozen trips from fire to boat – all at a fast gallop. The sweat was pouring off me. I had about three seconds of painting time each trip and the boat was beginning to look ridiculous.

I was applying, perhaps, the fifteenth lump when a dreadful roar went up behind me. I looked round at a sight that made my own hair match the spots on the boat. The oil drum was belching orange flame skyward with the noise and ferocity of an interplanetary rocket having lift-off. In blind panic I tried to run back with the brush but I was too late. It had stuck fast to the boat. I pulled with the strength of desperation and the handle came away in my hand. The hairs stuck up from the boat like some dreadful wart.

Bob and Douglas were moving backwards, their faces lit by the orange glare. Black smoke curled up through the trees. The oil drum was now red hot three quarters of the way up. The noise was frightening.

'My God!' said Bob, staring at me in horror. 'It's melted the handle off the can. There's a gallon of burning pitch in there.' The leaves on the oak trees about fifteen feet away were curling up like hedgehogs.

We threw ourselves onto our knees, digging up the earth with our bare hands and flinging it onto the inferno. The first six inches were mostly dead leaves and twigs and the whole lot ignited instantly and roared skyward, a pyrotechnical extravaganza in the fierce updraft. But with the strength of despair we were soon through to the peaty black soil and we worked like terriers digging out rabbits.

Ten minutes later we sat there in total exhaustion with sweat-streaked black faces like colliers who have just escaped some awful disaster. A great mound of black soil clicked and squeaked like a dying volcano, but the trees were safe.

It was some time before we turned wearily to the boat. It looked as if it had been struck by some unspeakable plague. The black sores were dreadful to behold, particularly where the brush hairs had stuck. We realised that if the owner ever saw what we had done to his boat we would be out on our ears. We struggled back into the boathouse with it and laid it on its side on the concrete rim. We gave it a push and it hit the water and settled slowly back onto the mud.

A crescent moon hung above the silhouette of black trees before we left. Apart from a few withered oak leaves all traces of our activity had gone. A little grebe called a watery trill from the dark lake as the car door slammed and the headlights picked out the overgrown path back to the road home.

I hate plastic and it is an unreasonable attitude. It is a material that, in its various forms, can solve so many problems of manufacture and convenience and even of efficiency, but there is something about its very nature that is repellent. Later that spring however, after the experience of tarring the wooden boat, I was prepared to compromise. I went to Southampton again and bought a fibreglass boat. It was called a pram dinghy and its shape was about as inspiring as its name. It was, however, so light and easy to handle that it could be picked up by one man and put onto the roof rack of a car with no trouble. It was light enough to hold over one's head and waltz round the yard. I know because I tried it several times.

There are strange aberrations of character which the psychologists tell us are derived from experiences of early childhood and I think it is true. Some of my own actions cannot, I like to think, be explained in any other way. My childhood was happy enough and the family lacked nothing of the essentials of life, but when one was born in the industrial north in the lean and hungry twenties one tends to feel that waste of any kind is wicked and that the purchase of unnecessary frills and flounces is anti-social. This probably explains why I bought the boat but refused to purchase any oars.

Like most optional extras, oars tend to be surprisingly pricey items and it seemed obvious to me that a broom handle with a rectangle of three-ply wood nailed on the end would serve the same purpose and show a commendable attitude to the correct use of the earth's resources. The shape of the boat was almost exactly like half an eggshell, cut lengthways, and when I put it on the water it scarcely appeared to dimple the surface. I lay on the concrete staging and looked underneath. Only about a square foot of the bottom was touching water. Why on earth would anyone want to build a heavy wooden boat again when this feather light shell would serve the same purpose? I was obviously in on one of the miracles of modern technology.

I stepped gingerly into the little craft with my home-made paddle and pushed off. The boat moved out into open water revolving round and round on some unseen axis. I jammed the paddle over the side to gain control then, aiming the bow at deeper water, I drove my paddle firmly backwards. The boat did at least six complete revolutions before I gathered my wits. I realised that I would have to push on both sides at once to make any progress but with a single paddle it was impossible. I settled for making quick jabs on each side alternately and jerked forward with the bow sweeping from side to side like a metronome.

As I have already mentioned, the lake was surrounded by dense woodland so that the wind hardly stirred its surface except from the

west where the water emptied out through the iron grill of a heavy sluice gate. Here there was a break in the trees and when the wind blew from that direction it was funnelled through with extra velocity. This happened to be a day when it was funnelling through.

The wind caught the hull as it came abreast of the overflow and, spinning like a whirligig beetle, it was swept along at an alarming speed up the centre of the lake towards the dense reed beds at the distant end. I abandoned all hope of controlled resistance and concentrated on survival.

To make matters worse, two swans were nesting up that end and I realised with horror that I was on a collision course with their nest. Dreadful stories I had heard ran through my head about how a single stroke from a mute swan's wing can snap a man's thigh bone like a carrot.

Many people who have stared straight into the face of doom tell us that on these occasions the mind will often clear suddenly and one can think quite calmly and rationally. It is true. I remember thinking 'How the hell am I going to control the boat when it is loaded with fishing tackle?' It must have been moving at about ten knots, still spinning, as we approached the swans. I could hear them hissing above the noise of the wind.

[65]

It must have been a frightening sight from their point of view too because their nerve seemed to give at the last minute and they thrashed the water into white foam before lifting into the breeze. They passed about five feet over my head as the boat swept under them and hit the nest. The bow lifted high out of the water as we shot up the mattress of reeds and I fell backwards with my feet in the air. I was lucky not to go over the stern into green water.

After several fruitless attempts to paddle back against the wind I gave up and sat it out in the reed beds. The wind slackened in about an hour and it took another half to work my way back to the boathouse. When I staggered ashore I was so dizzy I had to lean against the door jamb until the world became stable enough for me to walk in a straight line.

We all have to learn, but I didn't care for the smirk on the salesman's face when I went back to buy the oars.

In the summer, when the wild flowers are high on Rose Island, it is difficult to see the old wooden boat which I dumped there, but in the winter-time, when the trees are bare and the undergrowth has subsided, it emerges again and, now that I no longer have to spend tedious hours on its maintenance, it makes a pleasant enough background for the raft of tufted duck bobbing on the water.

Its place in the boathouse has now been taken by a pretty blue and white fibreglass boat which – in spite of my alarming experience all those years ago with the spinning dinghy – has proved ideal. It is a beautiful shape and, although it rides fairly high, it has a shallow keel which makes it reasonably easy to handle in a stiff breeze. It is always dry inside and needs no maintenance and very little cleaning provided that it is covered by a sheet when the swallows are in residence.

I am reluctant to admit to liking anything made of plastic but, like a new ping-pong ball, it has a certain delicate, shell-like quality and when it is drawn up onto the bank the kingfisher seems to like sitting on the oars while watching for small fry.

The Plank Bridge by the Pool   When the pool was excavated, fresh water came in through a short channel in the north-west corner. The stream was about eight to ten feet wide at this point and could not be crossed without some sort of bridge. I drove a pair of oak posts about eighteen inches apart down the banks on either side until they were firmly embedded in the bottom, then joined each pair together with a cross member level with the top of the banks. A single strong plank was laid across these in trestle fashion and wired firmly into place.

It is a strange and fascinating quality of water that it seems to make everything else associated with it as beautiful as itself. I have seen many strange looking ships and odd shaped boats but I have never seen an ugly one of either. All things that float naturally have a special visual beauty imparted by the water and it is only necessary for most things to be near to water to be transformed. The shoulder of a mountain reflected in a tranquil loch, a derelict warehouse by a neglected canal or a rough wooden post driven into a river bank or sun-dappled pond are all touched by this elusive magic.

The plank bridge was a temporary measure but within a few short weeks it had taken on a charm of its own. I found myself crossing and recrossing it for sheer pleasure. The broad plank reflected an umber stripe of ever changing shape on the water, the outline undulating on the gently moving surface. Rank grass and wild flowers closed in upon it, softening its hard edges and a patina of mosses formed above the water-line on the oak posts.

One morning I surprised a short-sighted water vole sitting on the bridge chewing noisily at a tender green shoot of yellow flag iris. Holding the morsel in clasped forepaws like a squirrel with an acorn, it was not aware of me until I was only six paces away. The tiny head stopped suddenly and the round furry body froze for a motionless second then flung itself like a lemming into the water. A loud plop and a target pattern of rings moved gently down the current as the vole, silver-grey now in a sheath of fur-trapped air bubbles, swam desperately under water until it dissolved into the blue reflection of the sky. The footbridge was accepted by the wild folk. It was a seal of approval and I was grateful.

Work on streams and rivers is always very hard. Work on a lake has different problems but it is just as exhausting. I know anglers and river keepers who declare that the joy of working with water is so fascinating that the physical effort is negligible. I envy

The plank bridge

them. I am forever going down to the waterside full of enthusiasm and determination to push forward with the latest plan. Wheelbarrow loaded with stone or wood, with saws, hammers, nails and most (but never quite all) the other tools required to do the job, I trundle along the banks to the point where the engineering project is to begin. I park my barrow at a strategic point and survey the site. I then go into a hypnotic trance. The beauty of the surroundings drives all enthusiasm for work away with the breeze.

When days are warm and drowsy, the waterside with its lush wild flowers and insect-droning air acts like a soporific drug. Down in the deep, clear water it is like a dream jungle. In the stillness of the pool the hornwort stands like great clumps of elms, dark and silent, with teeming shoals of tiny fish floating like black and silver birds amongst the branches. In the current of the stream fat clumps of bright green starwort and sinewy water buttercups wave endlessly like windblown tresses in a slow motion film. The sun sparkles on the ripples and the coots are busy in the reedbeds. There is always something more attractive to the mind and more acceptable to the body than hard physical work. A shoal of roach or chub drift into the clear transparent window of water surface immediately in front then flash away in panic and are gone beneath the bright reflection of the sky.

That tress of dark weed waving on the gravel bank is not ranunculus but the olive green and brown back of a big trout waiting for the endless supply of food particles brought to its mouth by the clear current. He looks two pounds at least. If I take careful note of his lie (two feet upstream from the yellow monkey musk on the far bank and a little left of stream centre) I might catch him when Bob and Douglas come fishing in the evening. It will be great kudos and something to boast about in the Mill Arms on Friday night. On the other hand, if I do succeed, that big brownie will not be there on the gravel the next sunny morning and something exquisite will have gone from the water. The theory that another trout will always occupy an empty lie is not true.

It was time I did something about the plank bridge. It was not really safe nor permanent enough. I went down for the third or fourth time to look at the job. It wasn't going to be easy. I sat on the plank and dangled my Wellingtons in the water. The gin-clear stream slowly cooled my feet even through the hot rubber and socks.

The lake looked lovely from this point as I basked in the hot sunshine. Perhaps it would be better to go back to the house and bring down my paints and easel. I had wanted to paint from this angle for a long time and, after all, it was my job. I had a living to earn and the bridge could wait until tomorrow . . .

On the other hand, the sun would be blinding on the white paper. The washes of colour would dry too quickly, leaving hard edges. The air was alive with insects and they would crawl on my face when I was holding the paint box and brush and unable to scratch my nose. They would drown themselves in my water jar and crawl in anguish across my painting, trailing cobalt or yellow in a crazy abstract of disaster.

Another problem of painting out of doors is that horseflies and bluebottles like burnt sienna. They swarm on it, struggling like metallic blue and green piglets on the sow. I recall a rhyme from childhood 'Green bice is not nice'. Burnt sienna is not nice either except apparently to flies. There was another rhyme that kids chanted over their first jumbo tin of paints –

> Little boy – box of paints –
> Licked the brush – joined the saints.

It doesn't seem to do flies any harm. They remain in rude health until I swipe them with my paint rag.

It was obviously the wrong kind of day for painting and even worse for building bridges. Tomorrow was bound to be better. Today was for dreaming. I got the rowing boat out and pushed off, left the oars where they were and just drifted.

Next summer the plank bridge was still there. More moss crusted the posts and thousands of tiny snails and water shrimps clustered on the wood below the water surface. A rich variety of small droppings showed clearly that it was a popular crossing place for wild creatures and it was often used in the early mornings by a heron as he surveyed the water and waited for careless fish to drift within striking distance.

I had been fishing one evening in the late summer with a doctor friend on the River Test. Some of the biggest trout are caught on dry fly quite late when the afterglow of sunset has almost drained from the sky. Being a man who enjoyed eating fish as well as catching them, he was well content with the fat trout he carried on a short length of alder twig. It was almost dark when we reached the plank bridge and glow-worms burned like discarded cigarette ends in the high rank growth of hemp agrimony and willowherb along the water's edge. I crossed the bridge and walked ahead towards the cottage. In a moment or two I realised that I could not hear his footsteps brushing the grass behind me. I turned round and saw him silhouetted against the western sky. He was in the centre of the plank, arms flailing like a demented windmill. His fly rod lashed and whistled as if he were at his first casting lesson.

Muttered curses floated on the scented air. I hurried back and pulled him to safety. In the dim light I could see from his face that I should not have laughed out loud, and that any hope I may have entertained of special medical treatment if ever I needed it had gone forever with the trout which had dropped into the water and was, even now, drifting quietly away in the dark stream.

A few days later I came within a hair's breadth of disaster when wheeling a heavily laden barrow across the plank. I reached the far end with about half the tyre width still on the wood and as we struck the bank the barrow tipped over and shot everything, including a big paper bag of nails, far and wide into the long grass.

The tubular metal handle of the barrow caught me a nasty blow in the groin which brought tears to my eyes. The thought of a visit to my doctor and the need to explain to him that I had lost my balance whilst crossing the plank made me wince even more than the blow, but luckily no serious damage was done.

When I did make a start on building a more permanent bridge it was still winter and any desire to drowse by the water's edge was quickly blown away by sharp winds. They roared across the valley with the noise of a hundred galloping horses and struck the plantation with savage blows, making the high poplars bend their heads in unison and send up complaining prayers like a tall thin congregation of darkly clad monks in an icy church.

The whole area where the stream enters the pool is peaty. The gravel lies far below at this point and the earth has the consistency of pudding. I had always wanted to have a little stone bridge over the water and had long ago realised that I would have to build it myself. This was the spot most suitable for such a feature and it was clearly now or never. A well built wooden bridge would have served just as well and would have been infinitely easier to construct but I would never have been content with it. I wanted to have one of those delightful little hump-back affairs that one comes across all over Britain in narrow leafy lanes where the road crosses a stream. They are beautiful in their simplicity and have a rugged charm quite unique to themselves.

It was clear that to make a strong, stable arch on which to build would be very difficult on ground where a solid foundation is hard to find. It would have required enormous quantities of stone and concrete below ground and even then it might sink a little and a cracked arch would be a disaster. The arch was the key to the whole operation and it would have to be very strong to support the weight of the stone and anything that might pass over it.

The problem was solved by buying a very slightly chipped but quite sound concrete sewer pipe and having it lowered into place on the bed of the stream. It was very heavy and quickly settled into the bottom and became quite stable.

Up in the yard there was still a lot of the stone left with which Bob and I had built the wall and steps and I was pleased to find that much of it consisted of big heavy pieces of the kind needed for the present job. (I was less thrilled about this when the time came for transporting it from the yard to the site!)

Filling in the ground between the sides of the pipe and the banks of the stream was rather tedious and very hard work but I had to ensure that this foundation was as solid as possible under the circumstances. Once this was done the hardest part of the work was the transporting of the heavy stones from the yard. Building up with stone and cement from water level seemed easy work by comparison, in spite of cutting winds and unwelcome freezing on occasions.

The family quickly noticed that when they were unwise enough to come down to watch me at work I would set up a whine about how the supply of stone was running out and how wonderful it would be if only someone would bring down fresh supplies. On occasion I had almost to burst into tears to make them push the empty barrow back up the hill to the yard, refill it with stones and trundle it back down again. Most of the time they did, of course, but if my back was turned they would deposit the loaded barrow

and run. As the work proceeded I had fewer and fewer spectators and before long only the inquisitive ducks dared to gather round and literally push their noses into my business. Jasmine, our black cat, well aware that she would not be expected to carry stones or cement, often kept me company for quite long periods. She would lie in the grass and watch me through amber slits until she saw a movement or heard the sound of a questing mouse in the grass. Then she would be alive in a second and, pressed close to the ground, would stalk the creature and leap upon some tussock with legs stretched stiffly before her. Sometimes she caught a shrew or bank vole and then, fearful of being reprimanded or even chased, she would streak away into the dense undergrowth of the plantation and I would not see her for an hour or more.

If the ducks were there when Jasmine came down she would not stay. If they came quacking up when she was with me she would immediately depart for she hated them. It is difficult to describe the look on the face of an embarrassed cat but the ducks embarrassed Jasmine. Whenever she went hunting along the water's edge they

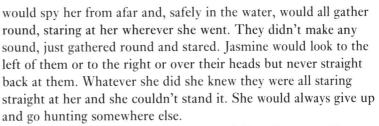

would spy her from afar and, safely in the water, would all gather round, staring at her wherever she went. They didn't make any sound, just gathered round and stared. Jasmine would look to the left of them or to the right or over their heads but never straight back at them. Whatever she did she knew they were all staring straight at her and she couldn't stand it. She would always give up and go hunting somewhere else.

When the bigger blocks of stone were laid and cemented into place the bridge began to take shape more quickly. If beauty is only skin deep then it shows how important the surface of everything must be. I have seen so many buildings erected by people far more skilled than I will ever be at structural engineering that are an affront to the eye. It is obvious that any bridge must first of all be safe but it should also blend with and enhance its surroundings, not look like a wound on the landscape. If the centre of my bridge was a sewer pipe there was no reason why it should continue to look like one when the job was finished. I built an arch of stones over each end of the pipe, overlapping the edges just enough to hide their mathematical precision and then built upwards and outwards from these central points.

The cement between the stonework was left rough and, where necessary, was deliberately roughened to produce a pleasant texture that would weather quickly and collect small pockets of moss and lichens. Here and there a small hole was left between the stones with a larger cavity inside. The entrance hole was made small enough to be of interest to the tit family and I was delighted to find one of these cavities occupied by a blue-tit the next spring. A number of gaps were left in the walls and a few handfuls of soil inserted as building continued. These spaces made ideal spots in which to plant house-leek, aubretia and other colourful rock plants when the bridge was finished.

The tops of the walls were finished off with flat stones which followed the gentle curve which I have always found so visually satisfying. The work was carried out mainly at weekends during the worst of the winter weather and often the wind was bitter and sleet came in the rain. It was hard and often uncomfortable work, but it was relieved by the animals and birds and by the sight and sound of running water and the changing colours and moods of the lake.

When the whole thing was complete and had been washed free of cement dust by the rain I mixed up some cow-dung with water in a bucket and threw it here and there about the stonework. Not only does this soften the harsh newness of the stone and cement but it quickly encourages the formation of mosses which give a mellow appearance to the surface.

[77]

There is nothing quite like the satisfaction of building something with your own hands and when it is something that is likely to be there for many years there is an even deeper pleasure. From the moment it was completed the little bridge has been a source of endless delight. It is a focal point for the eye, a perfect foil for the wildness of reeds and willows and an ideal place to sit and look at the view when the stone is sun-warm and hung with swags of pink, yellow and purple rock plants.

The water reflects a moving pattern of sunlight on the underside of the arch and the white ducks preen themselves on the rocks against the banks. Trout lie beneath it in the clear shaded water, particularly when the advancing year brings the instinct to move upstream. The heron leaves big arrowhead prints in the sand at the water's edge and the kingfisher flies down from his perch on the silver-green willow and lights on the parapet. He watches the small fry in the pool below and dives like a flash of neon blue to snatch a startled minnow from the shoal.

The hump-back bridge is another tender trap, another excuse to sit and dream and a perfect seat on which to do so.

One Man's Folly   In the spring of 1971 I decided to go ahead with the digging of the second stage of the lake. It had always been my intention to enlarge it once we had settled in and completed certain necessary work on the cottage. The first section of the pool had been dug out of the flat low-lying land at the bottom of the sloping lawn but I did not, at that time, own enough of this area to enlarge the water in that direction. Penny had reached the stage in her education where she had to devote a great deal of time to preparation for 'O' and 'A' Level exams and she had less and less opportunity to ride her pony. Tonda was a relatively young animal and needed plenty of exercise which he was not now getting. We were all reluctant to see him go but it would have been unfair to keep him in the changing circumstances.

His departure left Church Paddock unoccupied. This was the only area which could be used at that time to enlarge the lake, but there was a snag and it was a big one. The paddock, like the lawn, sloped steeply upwards from the water's edge to the cottage level and it was obviously going to be necessary to lower this land as much as ten or twelve feet just to get it down to water level where the real excavation would start. There was much head-shaking by everyone who knew about the idea and not a little good-natured leg-pulling by some who already considered my schemes bordered on extreme eccentricity, if not outright lunacy.

There was one important point, however, which few people seemed able to appreciate, but which I felt was very important and a heavy counter-balance in favour of the plan. It was that the change of level would be a great asset to the visual attraction of the water when it was all finished and landscaped.

The family knew that I would go ahead with it anyway because it was not by any means the first time I had pressed on with what had, at first, seemed a daunting undertaking and Bob, Douglas and Bill flexed their fingers for more plans to be drawn on beer mats.

One of the existing features which had to be sacrificed was the herbaceous border and fairly well grown hedge of cupressus trees. The wall on the south side of the stone steps would also have to go in order to get the maximum water area, but by planning the operation with care I was able to transfer the stone and most of the shrubs to the areas marked out for the two new islands I wanted and, by happy chance, it was this stone and the shrubs which gave the islands their names later.

There is a certain stage in land moving operations of any scale
when one begins to doubt the wisdom of one's own ideas. I must
admit that I had started to worry before we were down to water
level in the paddock because the volume of earth removed became
so enormous. The heaps of soil stood about twenty feet high and
looked like a range of black volcanic mountains. The whole
operation was economically possible only because about three-
quarters of the material removed turned out to be good quality
gravel and was, therefore, dug and carted away without charge to
me.

At a later stage I was very pleased to have the black mountains of
topsoil which had by then risen to bedroom window level, because
every bucketful was needed for the final landscaping of the banks.
The material was divided therefore into three sections: topsoil was
piled nearby for future use, gravel was loaded on to lorries and
carted away and the rest was spread over the other field to the north
and east of the cottage.

Men and machines were hard at work for many weeks but it was
all very interesting and exciting. When we were down to the correct
level I marked out the islands which were to be left with wooden
pegs, and the herbaceous border was simply dug up, bulbs, flowers
and shrubs, just as they were. The bulldozer took great buckets of it
straight out of the ground, trundled across to the island sites and
laid it down just as it was. The cupressus hedge was lifted in the
same way and deposited along the paddock fence. A proportion of
the trees, shrubs and flowers was lost in this rather drastic operation
but, on the whole, it was surprisingly successful and even if it had
been possible to do this work more carefully by hand it is doubtful
whether the results would have been better in the long term.

A section of our carefully built dry wall was lifted from the water's edge and the stone dumped on the marked site of the two new islands. I was not sure at this stage what I would use it for but decisions had to be made quickly before the islands became separated from the mainland.

When the earth had been removed to within a foot or two of water level, the high banks to the east and south were shaped. A path of about four feet in width was left to form a level walk along the water's edge. The digging of the water area then commenced and throughout the operation a causeway was left between the new excavation and the existing pool. We were quickly into the water table and, as with the earlier section, the area of water increased daily, but it was very muddy whenever the machines were at work and the causeway prevented the sediment from spreading into the mature section of the lake.

The bottom of the excavation was planned to provide wide areas of about fifteen feet depth where fish could find comfort in the long hot days of summer and protection from extreme cold in winter. Here and there shallow shelves were left near to the banks where dabbling ducks could feed and small fish find comparative safety from big trout and pike. The remaining area was varied in depth so that different varieties of water weeds and insects could find the conditions which suited them best.

At last we arrived at a situation where the new section of the lake was complete and its two new islands formed, but it was still separated from the original pool by the causeway. This was just wide enough for the digging machine to drive across in safety. The whole family gathered to watch at the great moment when the barrier was breached and the waters mingled into a single pool. The surface of the original lake was a few inches higher than the newly formed area of water, owing to the fact that the stream was running into it beneath the little stone bridge. The result was that, when the bank was broken, the clear water poured into the new pool and prevented any great surge of mud in the opposite direction.

As the excavator dug away the last of the causeway a certain amount of colour spread over the whole of the lake but the operation was completed in one day and the water began to clear quickly. Twenty-four hours later fish were rising over the whole surface and the new pool was already starting its maturing process.

At the western end of the old herbaceous border there was a group of silver birch trees and I was most anxious to preserve them. When the digging of the new pool was complete and the slopes smoothed out to form mowable lawn, the silver birches were left isolated on a little knoll. The side of this hillock facing the water,

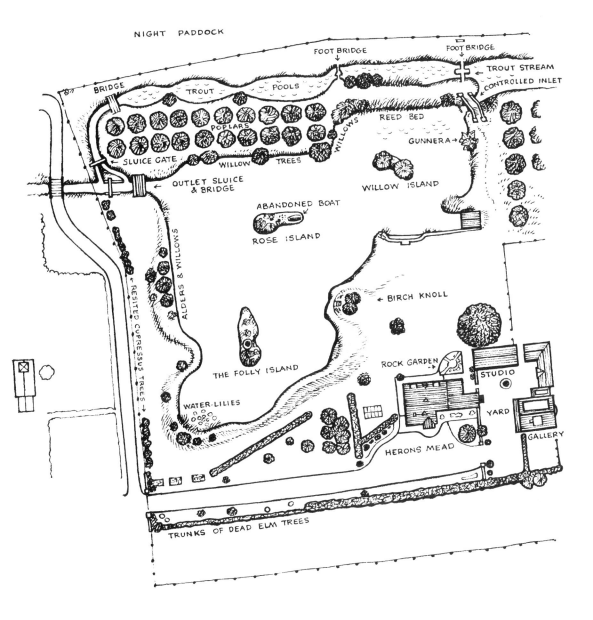

NIGHT PADDOCK

FOOT BRIDGE

FOOT BRIDGE

TROUT STREAM

CONTROLLED INLET

BRIDGE

TROUT POOLS

REED BED

POPLARS

WILLOWS

GUNNERA →

SLUICE GATE

WILLOW TREES

WILLOW ISLAND

OUTLET SLUICE & BRIDGE

ABANDONED BOAT

ROSE ISLAND

← BIRCH KNOLL

← RESITED CUPRESSUS TREES →

ALDERS & WILLOWS →

ROCK GARDEN →

STUDIO

THE FOLLY ISLAND

YARD

WATER-LILIES

HERONS MEAD

GALLERY

TRUNKS OF DEAD ELM TREES

The Birch Knoll

however, was so steep that there was a danger of the soil slipping down into the lake, particularly in rainy weather. Not only would this soon destroy the attractive shape of the promontory and deposit unwanted material in the pool, but it would quickly rob the tree roots of much-needed soil. Something had to be done fairly quickly before the erosion started.

It was for this reason that I decided to build a small terrace-cum-rock garden up the slope. I used stones as big and heavy as I could manage on my own and simply built upwards in a leisurely manner, allowing the shape to be dictated by the existing surface. The result was to produce three stone seats one above the other, with random rockery on either side. It was planted up with a variety of rock plants and miniature conifers and although it cannot be seen from the house it is a mass of colour in the spring and a visual point of interest from the other side of the lake. It is also a pleasant place to sit on summer evenings, protected from any breeze from the north or east, and if you stand on the top of the knoll by the birches you can see deep into the clear water below.

During the second week of May last year, my brother, Alan, was enjoying a concentrated effort to clear the pool of its many pike before the new flush of waterweed made spinning difficult. He had already taken about eight or nine fish during the week, but on the Thursday evening the light was already fading and he had had no success that day. He was casting towards Willow Island from the bank near the boathouse when I wandered up onto the knoll. As I looked down into the water from my vantage point between the birch trees, I saw a big fish in the water below. It was lying quite still on a patch of pale green weed and, although the light was fading quickly, I could tell from its size and general outline that it was a pike.

Through a series of nerve endings along the lateral line which passes from gill cover to tail on both sides of their bodies, fish are highly sensitive to the smallest vibration. Footsteps on the bank can send them dashing for cover in panic but they cannot hear the sound of a human voice. I called to Alan that I could see a pike and suggested that he move cautiously to a point about forty feet to my right and cast diagonally across towards one of the new islands. He could not see the fish but I could see his spinner flashing below the surface as he retrieved it. As the bright metal lure passed about two feet in front of the fish's nose, it lunged towards the bait but missed. Too late to catch it without a chase, it circled back to its original position and lay still. I explained to my brother what had happened and he cast again on the same line. This time the pike was alert and took the approaching bait in a curving slash of

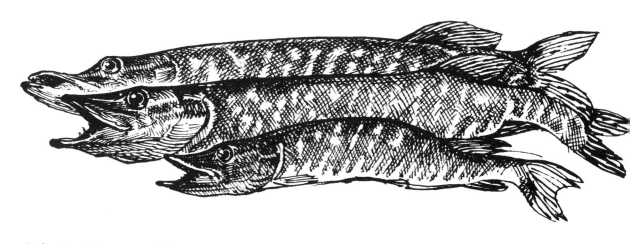

Jack pike taken from lake march

incredible speed. He made the placid water surface boil and rock from bank to bank before he was beached down by the steps and, at just over ten pounds, he proved to be the best pike taken all week.

There were now three islands. Willow Island was already well established and I turned my attention to the two new ones.

I was not quite sure what I wanted to do with the stones which had been dumped on them but I had a vague feeling that I might form some sort of landing place which would make it easier to get ashore from the boat.

I started by building stone steps in the little bay of the island near to the church. When they were done they had a slightly lost appearance, leading nowhere, and as there was still a lot of stone lying about I started to build somewhere for them to go. I have always been attracted to the great eccentrics of the past who built follies on their land. I know why they did it, for the uselessness of these buildings is the essence of their charm and I could think of no good reason why I should not indulge a monumental – if modest – whim of my own. The first ring of stones seemed to cause great amusement and there was much witty speculation as to what I was building: a gazebo? – an outside privy for sailors? – an outlet for an underground escape tunnel in case a domestic fracas should get out of hand? – and many more.

I put a little doorway by the steps for no other reason than to puzzle the comedians more. As the tower rose a tiny window facing the church caused a buzz of speculation. The wall rose another foot or so then stopped. There was more stone on the ground and the top of the wall was not complete but nothing more has been done since. I like it the way it is. The shrubs have closed in upon it now and as the alders and birches grow it becomes less noticeable, but the birds seem to like it and although it is literally a pile of loose rocks with no mortar to bind them, not a single stone has moved in the winter gales since it was abandoned. It doesn't matter if a few do fall. No one will notice an extra rock or two amongst those on the ground. It has always been referred to as the Folly Island.

The steps are never used, except by waterhens to build their nests upon, for they are far more hazardous as a landing place than the pebble beach at the far end. Both mallard and call ducks have nested on the island from time to time beneath the mock orange or under the forsythia bush, but they never go inside the tower for they know it for what it is – just a folly, and they refuse to take it seriously.

The centre island is the haunt of butterflies in the summer, for among the shrubs which were bequeathed to it from the old border is a buddleia. Its mauve panicles are made more brilliant by adonis blues, brimstones, red admirals, peacocks and many others. Pink roses grow there too among the tangle of branches and give it its name – Rose Island. The stones from the old wall still lie where they were dropped, for I have never got around to using them yet. Perhaps one day when the sun is warm and the mood is right I will start the jetty which I vaguely have in mind, but it will have to be when there are no nests of ducks or coots or moorhens tucked between the stones.

The trees on Willow Island were tiny saplings when I first planted them but are now big and strong enough to be climbed in perfect safety and in high summer they spread a leafy green dome over the whole island and trail their fingers in the water. Many tufted ducklings have been hatched there along with coots and waterhens, but the mallard do not choose it for nesting sites and only one white duck has built on it and, although she was successful in hatching six delightful yellow ducklings and four mallard ducklings, she has not nested there since.

In the spring of this year a coot built her nest and laid a clutch of eggs. A pair of Canada geese had been visiting the lake rather frequently and began to cast envious eyes on the property but were driven off after a short, noisy battle. A good deal of water flew about but the geese beat a retreat that was rather undignified, for they are the biggest geese in the world and they made the coots look as insignificant as irate blackbirds.

The female coot was sitting on her eggs when the Canadas returned the next day and began to build their own nest about twelve feet away. The male coot bustled angrily about, doing his best to defend his territory, but he seemed to lose heart before his giant adversary. His mate sat bravely on her nest as the situation grew tense.

In mid-afternoon I was in the yard talking to David when we heard violent splashing from the pool and rushed round the studio to see what the noise was all about. The male Canada goose was thrashing the water into foam against the wooden piles of the island and plunging forward repeatedly with his black neck. When his frenzy subsided the male coot's lifeless body was floating with the head below the water. His mate still sat on her eggs only six feet away.

The next day I picked the floating coot from the water and buried him with a feeling of considerable respect. There was nothing more he could have given in defence of his mate and her eggs. The goose swam menacingly near to her but she did not flinch all day. In the morning she had gone.

I didn't think she had been killed because there was no sign of her on the island or in the water. It was the first time that the geese had nested on the lake but it was soon obvious that they did not approve of neighbours.

About two weeks later the grass and wildflowers had grown quite considerably and I noticed the female coot returning to the island. She was half hidden by the new growth but was sitting again on her old nest site.

For reasons best known to themselves, the Canada geese made no further attempt to molest her, although the male continued to harass all the other water birds from time to time. He spent his days swimming slowly up and down near to the island or preening himself on the stone steps nearby; then for no apparent reason he would take off, fly across the water and attack some innocent duck or waterhen feeding quietly on the far side of the pool. On several occasions I saw him leave the lake altogether and fly across Night Paddock to chase birds which had been feeding two or three hundred yards away.

One morning when the goose had been sitting for about a month I noticed she had left the nest and was standing on the bank near the great rhubarb-like gunnera plant which was uncurling its big spiky leaves in the warm sunshine. The gander was with her and, although he gave me a fierce stare as I approached, he seemed a little more nervous than usual. The goose moved to the bank side and slid into the water. Five beautiful grey goslings emerged from the grass and followed her. The gander stood his ground until they were all safely afloat and then, with ponderous dignity, he moved slowly down the bank and followed them.

They were a handsome pair swimming about the pool or loafing among the white daisies on the lawn with their pretty goslings around them. They stayed with us for five more days before I saw them one afternoon crossing the mistletoe pool on the way to the river and wide world beyond. The little lake had added five wild geese to the impressive number of birds that had first opened their eyes beside its bright water.

For a day or two after their departure the pool seemed unusually quiet but soon more mallard returned and call ducks and tufted came back home. A little grebe built her floating nest in the reeds by the boathouse and I almost trod on a mallard sitting on a clutch of eggs beneath the nettles in the plantation. Our distinguished guests had departed and the pool was back to normal once more. They had not gone more than a week when three baby coots emerged from the nest site which had been defended so gallantly by the little black and grey bird they had never seen.

Little grebe by boat house

The Trout Stream The limited flow of water which came in under the hump-back bridge was generally fairly clear but it was depositing a certain amount of silt in the lake and there was not enough flow to carry it out again at the other end. I had decided from the start to bring a supply of water from the carrier as soon as it was practical and, when the second stage of the pool was complete, it seemed an appropriate time. It was obvious, however, that if the new supply was to run directly into the deep water it would deposit even more silt which would cause serious problems in a few years' time. For this reason I decided to make the new stream by-pass the pond and arranged a sluice near to the bridge so that water could be taken in when it was needed in the hot drier days of summer and excluded when the streams ran deep and coloured in the long wet months of winter.

A sluice was built on the carrier near to the droopy bridge where the salmon spawned in the autumn shallows and the water was directed down the ditch towards the lake. The ditch itself was then widened and deepened, leaving narrower channels here and there to speed up and vary the flow. It was planted with water buttercup, starwort and a variety of other weeds which would encourage and shelter the abundant insect life that feed the sleek and wily trout. Clumps of reeds and wild iris were planted here and there along the water's edge and willow saplings to give some shade to the water and interest and variety to the banks. Watercress and clumps of forget-me-not like a pale blue mist appeared without my aid. Wooden footbridges were built across the narrower sections to make access easier for maintenance and fishing, for the stream is about a quarter of a mile long from the sluicegate on the carrier to the point at which it empties out beyond the lake and runs away through my neighbour's garden into the old canal. Half a mile further on it rejoins the carrier and thence the water returns to the main river.

Because of the speed at which all water plants and trees grow, it needed only a couple of seasons to turn the once turgid ditch into a delightful trout stream. It soon became a favourite place for Bob and Douglas to meet me in the long summer evenings and we often fish on until the moon is high and the rising trout are more easily located by sound than by sight.

Bob and Douglas put fifty good trout of about one and a half pounds into the stream and I added a hundred fingerlings. The smaller trout were put in partly to grow on into bigger fish for later years and partly in the hope that they would divert the attention of herons, cormorants and other predators from the bigger trout. We are not purist anglers and enjoy catching the occasional roach, chub and grayling but our little stream, as well as the pool itself, is constantly menaced by pike. These voracious creatures move into our territory from the main river in large numbers and certainly enjoy more trout than we do.

Having too many pike in a trout stream or pool is a serious problem. They will destroy the whole population in a few seasons if not kept in check. Even the smaller pike are a danger to the trout but we regularly find ten- or twelve-pounders lurking in the shallows and quiet backwaters where it is almost impossible to catch them with rod and line. Apart from the fish which are killed by the pike, I am sure that many more are simply frightened or chased out of the streams into the main river where they have more chance of escaping from determined predators. It is a constant dilemma deciding how to keep a fishery in sensible balance without waging all-out war on certain creatures that, under natural conditions, would have a valuable part to play in the balanced order of things.

We catch pike frequently, sometimes four or five in an hour, on spinners or Devon minnows and, when they are in more comatose mood, by slipping a noose over them as they lie in ambush among the bankside reeds. On one occasion, more by good luck than good management, I succeeded in flinging a huge specimen onto the bank with a garden fork. He had been lying beneath a small wooden footbridge in a narrow and very shallow side ditch. It was impossible to get a spinner near to him and I never use live bait. In the circumstances of the late evening and the lack of time to work out more subtle methods, a garden fork was the only weapon available to me.

There seemed no chance whatever of getting him out, for he could see me as clearly as I could see him. He was directly beneath me as I lay flat on the wooden slats and leaned over with the fork. My wife, who was with me at the time, sat on me to prevent me sliding head first into the water. When I reached down I could not understand why he remained there. He had a clear run of only about five feet into the deeper water of the stream yet, apart from a gentle rippling of his pectoral fins and rounded tail, he lay as still as a stranded log. He was no more than three feet below me, almost touching a big clump of reeds on his left flank. I counted to three to prepare my wife for the big effort and swept the fork as hard as I could towards the reeds.

The next few seconds were a blur of confusion and flying muddy

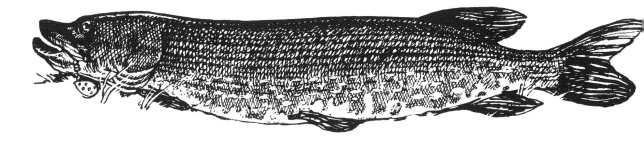

water. As the fork shot him sideways against the dense reeds, they bent towards the bank and formed a smooth ramp. He simply flew up the slope onto the grass. My wife and I scrambled up and got between the thrashing monster and the edge of the water.

When we recovered from the battle which followed we weighed him on our biggest spring balance. He was just an ounce or two over twenty-two pounds. That one pike was capable of swallowing our largest trout as easily as a gourmet might take down an oyster. As pike go he was hardly big enough to approach the record class but there cannot be many fish of his size that have been taken on a garden fork.

Like cormorants, pike have a sinister, rather prehistoric appearance. They would look quite at home with the Pteronoden and Ichthyosaurus in the primeval swamps. I am sure that some of our pretty ducklings fall victim to the bigger pike, as if they were not already quite capable of finding a hundred ways of committing suicide without the help of enemies from above and below the water.

From time to time we see eels both in the stream and in the lake and many mornings I have watched a heron struggling to subdue a lively specimen before swallowing him head first. I am surprised and delighted that I have never yet caught an eel when fishing on my own water and I hope I never will. I once had an experience that put me off these creatures for all time.

One of the perks of being a freelance artist is that one can occasionally take a little time off in the middle of the week. One sunny afternoon when Douglas, Bob and I had the little lake where we tried to tar the bottom of the boat, I looked out of the studio window and knew that I could not bear to stay indoors any longer. Douglas was at work, I knew, but there was a slim chance that Bob might desert his pigs for an hour or two so I rang him up.

'Go fishing?' he said in answer to my enquiry. 'Look mate, I've got three hundred pigs to feed, half a ton of muck to shift and a sow about to farrow any minute. Sorry, I can't make it. Enjoy yourself – you lucky devil.'

The lake was dead calm and lay like a slightly dusty mirror reflecting the surrounding trees and every detail of the wispy white clouds and blue sky. We had seen some fairly big trout rising on odd occasions in the late evenings, certainly much bigger than the rainbows we had put in, and it was these fish I was hoping to contact. I anchored the boat in deep water with a heavy piece of iron piping on a long rope and decided to try a wet fly sunk deep. My landing net lay in the bottom of the boat and my fishing bag, full of the glittering jumble of angler's toys, lay open against a heavy saucepan of trout pellets which I intended to scatter on the water before I went home.

The dense trees enclosed the water like protective arms and held it secret and remote from the outside world like some watery Shangri-La. The air was hot and silent. No bird sang in the woodland, no coot stirred in the reed beds which stood motionless in the heavy atmosphere. Patches of bright green waterweed glittered in the sunshine against the black reflection of the oaks.

There was a strange eerie loneliness and a silence so intense that my eyelids drooped as I gently agitated the artificial fly deep in the water. The slender rod tip swayed. It was the only movement in sight. I felt a genuine pang of sorrow for anyone who had to work on such a day.

There was a slight resistance on the line – then nothing. It was there again and the rod tip bent slowly into an arc. I gripped the butt tightly and struck. I knew I was onto something interesting when the boat started to move along the surface until the anchor rope tightened and held it, swaying from side to side. Wide awake in a second, I began to play the fish. Anyone who has never experienced the thrill of having a lively fish on the end of a willowy light fly rod will never understand the breathless excitement of such moments. The rod bent violently and line screamed off the reel as the boat was pulled round in circles at each change of direction the fish made. My mouth was dry with the concentrated effort of keeping the line away from the anchor rope. With only three-pound breaking strain filament holding the fly I dared not resist the powerful rushes too strongly for fear of breaking such light tackle.

It must have been about eight minutes of tense excitement before the surface boiled and I caught the first glimpse of my prize. It was an enormous eel. The last thing on earth I wanted to see at that moment was an eel.

Instinct told me that my gear would snap in seconds if I didn't get my net underneath the creature quickly. Holding the curving rod high with my right hand, I grabbed the net with my left and scooped under the thrashing, writhing turmoil and heaved. The weight was in the net – then it had gone. The rod bent again and line snaked out from the reel but this time it passed through the net on its way down. The wretched creature had slipped through the mesh and left a glutinous mass of grey jelly in the net.

If I let go of the landing net it would certainly be pulled into the depths by the line. If I held on to it I had no hand free to control the flying reel. I decided to sacrifice the fly and trout cast and, heaving the net inboard, I put my foot on it and reeled in line as hard as I could. With a trout on the end the cast would certainly have broken. I expected the line to come free any second – but it didn't. In a few moments the monster was lashing the surface once more by the side of the boat and rings pulsed out across the pool from bank to bank setting the lily pads rocking.

I could not reach the scissors in my fishing bag without taking my foot off the handle of the net. I would have to risk all with a great heave. Once it was free of the water the increased weight of the thrashing eel would snap the nylon filament like cotton. Holding the reel with my finger, I took hold of the net, pushed it over the side and gave a great upward jerk with both arms.

The creature flew out of the water and hit me in the face. It slid down my jacket and writhed on my lap before slipping between my knees on to the floor. I was covered from head to foot in slime. For a few seconds I was rigid with shock and horror.

In a blind panic I grabbed the only weapon to hand. Trout pellets flew in every direction as I swung the saucepan above my head and lashed out. The pan struck the fibreglass bottom with a deafening crash. I struck and struck in all directions. The bangs reverberated back and forth from the surrounding trees. Water fowl screamed in the reed beds and woodpigeons clapped away in panic. I never touched the eel once. He was into my fishing bag and out again scattering spinners, fly boxes and a welter of angling gear all over the boat. One blow from the pan flattened a delicate fly box into a single sheet of aluminium. The rod, which was bending and twisting like the eel, suddenly flicked straight and lay still. The eel was free.

He shot under the seat into the bow. I scooped at him madly with the pan and to my unspeakable relief he flew over the side and disappeared. I sat there for a time breathing heavily and surveying the chaos, my mind numb with the uproar. At last I pulled up the anchor and rowed back to the boathouse. There were some clean rags in the car boot and I set about cleaning up. The slime had set over everything including my hands and face like dirty rubber solution.

The afternoon was far gone by the time I had finished and I did not fancy sinking any more lures down where that monster lurked. I went home, cleaned up my clothing and had a bath. I was having tea when the 'phone rang.

'You're back then are you, you lucky devil,' said Bob. 'Did you pick up anything interesting?'

Our fishing on the little stream is almost entirely restricted to dry fly for trout and big spinners, plugs or Devons for pike, so we are, happily, unlikely to attract eels and I am very thankful for it.

At the point where the stream turns sharply to the left and empties out beyond the outflow sluice of the pool the ditch is deep and narrow. The water flows very fast here and it is quite impossible to cast a dry fly in the restricted space. It is a favourite haunt of big trout however, for they like the fast, cold water and feel safe in the green shadows of the weed tresses. This is the best place to start fishing the evening rise when the blue winged olives are hatching, and if you throw a few trout pellets into the small pools upstream the big fellows can be enticed out into the open water where they then seem content to stay and feed on surface flies. The water is so clear and the stream so small compared to the main river that much more care has to be taken in concealing oneself to avoid putting the trout down for long periods. There are good chub here too and often a shoal of roach. The three small pools have produced some fine trout in their time, often from two to three pounds in weight and there is no doubt that, while our own trout are free to swim away both up and down stream, other trout are equally able to swim into our stretch, so that in general the stock seems to balance out over the seasons. The next pool upstream of these three has a thick reed bed on one side and willow bushes which make casting a ticklish business beyond the footbridge, but the upper end where the water runs in through two narrow channels is deep and fast with great swags of swaying weed on a gravel bottom. It is a favourite place for trout, chub, roach and the lurking heron.

When fishing here one evening in 1974 I noticed my son, David, near the plantation waving his arms to attract my attention and pointing to something high above Night Paddock. I followed his directions into the bright blue dome of the sky and saw a great bird flying with slow wingbeats up the valley. It held something in its talons but I could not make out the detail without my binoculars. David had his glasses with him as usual and was rewarded with a clear view of an osprey carrying a fish as it flew north.

Beyond the two narrow channels close beside the hump-back bridge is a long straight stretch of water. It is a good place to fish late on a summer night when the sun has gone down, the sky is fiery red on the western horizon and the cool disc of the moon is reflected in the quiet water.

It is often too dark to see the ring of a rising fish, but the sound of flies being gulped down gives away their general position and it is possible to manoeuvre quietly until the fish or, rather, the rings they make are picked out in the long rippling reflection of silver moonlight.

Fishing for these trout is a very special and delicate technique. It consists of casting a foot or so upstream of the silver reflection and floating the fly downstream over the rising fish. If the cast is accurate the line dents the surface film just enough to pick up a flicker of light along its length and the fly itself is discernible as a tiny star at the far end. It is surprising how accurately the dry fly can be cast in these conditions and the fish when they are caught are almost always big trout. Bob is particularly skilful at this technique and I have known him take the best fish of the day on many occasions even when there was no moon and he was casting only to sound.

Blue Bridge

At the end of this stretch the water turns at 90° to the left under the blue footbridge and on to Mistletoe Bridge, where three tall poplars stand with big bunches of Christmas decorations growing between their branches. I watched two cuckoos chase and tumble and swoop about these trees in mating flight for a full ten minutes one May afternoon but they paid no special attention to the mistletoe boughs. At this point the water surges through two big pipes beneath a drainage ditch into a wide pool where fish seem to gather, as if in some underwater market place, but they are not easy to catch. They seem to be extra wary where the water is wide open and spend much time feeding on the rich insect life which comes into the deeper water through the siphon pipes.

From here on the stream narrows into a string of pools which continue under three small footbridges up to the main sluice gate on the carrier bank.

This is a particularly interesting and beautiful spot, for below the sluice the water has scoured out a broad deep pool. The surface is never still when the gates are open and the fish which gather here can only be seen when the sluice is closed and the surface settles to a clear glassy window. The water is so crystal clear that it is necessary to crouch low behind the yellow loosestrife and hemp agrimony and peep carefully through the vegetation to see the shoals circling in the unaccustomed stillness of the pool.

Up above is the carrier itself, gliding smoothly along to the slender droopy footbridge on the right where the water breaks into fast sparkling wavelets as it rushes past the iron grill of the sluice

The droopy bridge - april

gate and on down to the cattle bridge at Lime Tree Drove. This is a wonderful place for both trout and the grayling which love the clear tumbling water, and it is a favourite spawning place for salmon. There are almost always some salmon here in the autumn and four years ago three pairs of fish cut their reds in the bright gravel below the bridge.

When I was working on the bank of the carrier here one Saturday in the long hot summer of 1976 a water vole, surely the most timid and nervous of waterside creatures, emerged from the bank directly below the length of timber I was hammering into place. She had a tiny baby vole in her mouth and she swam downstream within a few feet of my pounding hammer. To my astonishment she returned a few minutes later and collected another baby from the watery recesses of the bank. There was no way in which she could rescue her family without this total exposure but she returned four times and carried off a minute offspring on each trip. The total devotion of bird and animal mothers when their young are threatened is an endless source of wonder.

Twice I have surprised a delicate roe deer there by the footbridge and watched it crash away into the tall dry reeds, cross the stream and disappear into the withy beds on the other side.

From here on, in both directions along the carrier, is some of the finest dry fly fishing water in the world; far better than my little stream can ever offer, however much love and attention I give it. But time and again I find myself turning back along the little waterway where the hare and partridge lie close in the high tussocks of summer and the fierce December stoat on the footbridge sits upright to look me over before disappearing like a brown and cream snake beneath the frost-hung briars.

The distance is short from the carrier bank to the pool by the hump-back bridge and not much further across the little lake to the far bank beyond Folly Island, but it measures a whole world of beauty and delight, which ranges in scale from the deer in the carrier reeds to the bright rustling dragonfly over the water lilies below the church bank.

The Great Bank Mystery

Fishing down the stream one June day I came upon a big hole in the bank about three feet deep and nearly as wide at the top. The earth was scattered about as if a small bomb had exploded and it needed several barrow loads of extra soil to fill it in again.

It was a mystery, for I could not imagine any wild animal digging a hole of that size and shape. A week or so later I came upon another excavation of the same kind. It was weird. The holes seemed to have no purpose and were certainly not the work of rabbits, badgers or foxes. No one could suggest a satisfactory explanation. From then on, I kept a sharp eye on the area but saw nothing which might explain the phenomenon.

Some weeks later, when I had almost forgotten the incident, I noticed a herd of cows against the fence watching something on the bank of my stream. Keeping out of sight as much as possible I crept through the plantation until I could get a clear view and saw the back legs and tail of a big dog. His head and most of his body were down a hole from which showers of soil were shooting in lively jets. The animal was so busy at his task that I was within feet of him when I shouted. He was a red setter and he emerged from the excavation with wet soil and a slightly surprised look on his face. He regarded me with a rather silly grin, panted a few times and disappeared down the hole again. More soil flew out.

Taking a deep breath I leaned forward and barked loudly at him. He leapt from the hole, shot through the fence and went across the fields as if the devil were on his heels. I concluded that he was a dog with a highly sensitive nose and a keen interest in water voles and the mystery was solved.

I was taught how to bark at dogs by my father who took a special pride in it, but it should be done sparingly and with discretion. When my brother and I were very young my father, who had been a town child as we were, would regale us with stories of how his father had been known far and wide in New Ferry for his ability to manage animals, however intractable. Savage dogs were his speciality. Grandfather, it seemed, had a penchant for bull terriers and other aggressive members of the canine tribe. He kept one always in the yard, attached to the kennel by a chain, but it was rarely the same animal for more than a few weeks at a time. He swopped dogs like other people swop foreign stamps and his children (a round dozen if I remember rightly) were never quite sure whether the current occupant of the yard was friend or foe.

They took to gathering at the window when they came home from school and trying to assess the degree of risk involved in going down the yard. The snag was that the yard was narrow, the chain long and the lavatory down the far end. When the dog was of uncertain temper, which was frequently, the family had to go out of the front door, round the block and in through the back yard door to get to the lavatory.

The barking incident, however, concerned a period later in my father's life. He walked to work about three or four miles every morning at about six-thirty summer and winter, and he had for some time been menaced by a dog that lived in a house with no front gate. The animal's technique, it seemed, was to lie in ambush just inside the gate posts and when father drew level he would leap out at him with blood-curdling yelps. This practice, particularly on dark mornings, was, understandably, a bit unnerving.

'I cured *him*,' said father. 'No bother.' We begged to be told his secret. 'You have to let dogs know who's master,' he said.

'But what did you *do*? How did you show him who was master?' we squeaked.

'Easy enough,' he said. 'I waited until he flew out barking and then dropped on all fours and barked back. I've never seen or heard him since,' he added.

It was very impressive. I was in my formative years. All my father's stories of dogs, or even cats for that matter, were full of delinquency and menace and whenever I met a dog in the street I tended to feel that it was either him or me. Father's 'surprise' technique gave me new confidence and I took to barking at dogs whenever I met them. It worked like a charm. The look of surprise and terror on the faces of the local dogs made me drunk with power.

I was in my sixth year, head down and hands deep among the conkers in the pockets of my short pants. As I rounded the corner by Stewart's greengrocery shop I met an airedale. He was a big, well-built animal. He was, naturally enough, walking on four legs whilst I was upright but our eyes were on exactly the same level. I can still see the crinkled texture of his big wet nose. We were both a little taken aback by the sudden encounter but my reflexes were fast for my age.

I barked in his face.

The next few seconds were, without doubt, the most terrifying of my life up to that point. The hair still stands up on the back of my neck when I recall the dreadful cavity of his open mouth, the ranks of yellowish teeth and the violence of his bark which almost shattered my eardrums.

My father had never explained the next move in such circumstances but instinct supplied the answer. I fled back up Park Road with the airedale in hot pursuit. The end must have been quick and painful had not fate in the form of a black cat stepped in, or rather run out, from the back entry. I do not know what happened to the cat – or to the dog for that matter, for I never saw either of them again – but I fell in our back yard door with my confidence permanently damaged, though luckily with all my conkers still intact.

To the best of my knowledge I have never barked at a dog from that day until I frightened the red setter and, if I had stopped to think about the airedale, I doubt whether I would have taken the risk. But the skill must have remained with me over the years for the dog has either given up digging out voles or has decided to pursue his interest elsewhere.

Swan Song   Almost every day, as I work here in my studio or walk the banks of
the pool and river, I hear the powerful, romantic music of the mute
swans' wings as they fly overhead or plane down, with black webs
stretched before them, to taxi onto the water. It is impossible not to
be impressed by the muscular symmetry of these great airborne
creatures, or to be moved by the visual beauty of their white flight
against the blue-grey of distant woods. But I have never been able
to love them. They will remain forever mean, merciless and
magnificent.

A pair of swans came to the pool in January of the first winter
after it was excavated. Like all swans on water they looked very
beautiful and they cruised up and down in regal fashion with the air
of owning the whole place absolutely. Although they fought
spectacular watery battles with others of their kind who arrived
later, they showed no inclination to attack my white call ducks or
the little raft of plump tufted duck that had settled in for the
winter. They worried me a bit, however, for several reasons: firstly,
I could not forget the destruction of the ducklings so long ago and
so far away; secondly, they spent much time loafing and grazing on
the lawn and in a short time had fouled a large area as if it had been
used by half-a-dozen big dogs and, although they tended to move
slowly and resentfully into the water if I walked down the lawn,
they made it quite clear by the thickening of their feathered necks
and by loud harsh croaks that they regarded me as an unwelcome
interloper.

I was looking forward with great excitement and pleasure to the
spring and hoping for at least some ducklings from both the call
ducks and wild tufted ducks; so that when, after some weeks of
occupation, the swans showed some signs of collecting material for a
nest, I decided that they must find other quarters.

I had a very long piece of soft white rope and my wife or
sometimes my son helped me by taking hold of one end and walking
round to the other side of the pool with it. We would then walk
along opposite banks together with the rope swinging between us a
few feet above the water surface. Only the first stage of the pool had
been excavated at this time with one island. This was to become
Willow Island, but the saplings were then so small that they did not
interfere much with the rope which we simply flicked over them.

The swans retreated before us but showed little sign of great
concern. On several occasions when we had cut them off in the

narrow end of the water they beat their great wings on the surface
and took off, rose above the rope and splashed back onto the
surface on the other side. The whole operation had then to be
repeated; on several occasions four or five times before they became
concerned about misjudging the length of runway and flew away
across the valley. Our triumph was short-lived, however, for they
were always back when we looked out of the bedroom window the
next morning.

They were obviously determined to stay and I was not happy
with the situation.

I had been fishing on the River Test for some years before this
and had been astonished at the aggressive behaviour of swans once
they had a nest with eggs. Whole sections of fishing bank, perhaps
fifty to a hundred yards long, were useless where swans were
nesting, for the cob would threaten anyone moving along the bank
in that direction and would not hesitate to attack if he thought the
eggs might be threatened.

I was forced to admire the courage of these magnificent birds but
I did not fancy being attacked on my own lawn or exiled from my
own water. Nor did I look forward to the possible destruction of the
ducklings which I wanted so much. I was frankly baffled but, as so
often happens on such occasions, the answer came simply when I
was least expecting it.

Looking through the bedroom window one morning when the swans had been in occupation for some weeks, I saw a heron standing on the island attempting to swallow a fish. A heron's sight is incredibly keen over both long and short distances and he must have seen a movement at the window for he dropped the fish and lifted his lank body into the air with slow beats of his broad grey wings. He turned towards the gap in the plantation trailing his long legs, then, lifting them up so that they stuck out straight behind him and drawing back his head until it rested close to his body on his folded neck, he swung away across the valley towards the main river. The fish flicked a few times on the island grass and lay still.

When I had first put the trout in the pool I had expected to lose a fair proportion to predators of one kind and another and it seemed a reasonable exchange for the pleasure of having all kinds of wild creatures living as neighbours, but I was not anxious to see my fish plundered simply to die on the bank. I dressed quickly and was down to the boathouse in minutes. I had not pulled more than two or three strokes away from the boathouse when I was startled by a great noise of thrashing wings on water. The swans cut a long sparkling wake of turbulent water before lifting into the air and following the heron's path through the poplars out into the valley.

The trout had not moved on the island although he was not yet quite lifeless when I touched him. Blood oozed slowly from a gigantic wound in his side where the heron's bill had transfixed him. Another tear on his opposite flank showed where the mandible had passed clean through. It was one of the many tragedies of waste that occur constantly where nature is allowed to take its relentless course.

The swans, meanwhile, did not return for a whole week and when they did I went down to the boathouse again and rowed slowly towards them. Evidently it was my intrusion on their natural element which upset them for within a minute they were half a mile away across the valley and did not return that year.

Swans have often visited the lake since then and now I do not disturb them for I know that they can easily be sent on their way without fuss if they decide to nest. I have had reason to do so on two occasions since, when they have started to collect reeds on the beach of Folly Island. Mostly they feed on the lush weed growth of the lake for two or three days and then depart as suddenly as they came.

Swans have a vaguely medieval quality as far as I am concerned. Perhaps it is that I tend to connect them, along with boars' heads, with pictures in story books of great kingly banquets and junketings. It is strange how food seems to run in fashions. In the middle

ages roasted swan and pike were both considered great delicacies but few people think of either nowadays as a source of food.

When my daughter, Penny, was a teenager and crusading against all kinds of food wastage – and very commendable too – we were cajoled and bullied into eating pike which I frequently caught both in the lake, stream and river. I found it quite pleasant to eat – a bit like fish and potato cakes – quite mild in flavour and not in the least earthy, as so many people imagine it to be. Considering that pike are flesh-eating fish it is difficult to understand why so many people imagine that they taste of mud. The rest of the family declared that they could take it or leave it. For one reason or another it is true that relatively few people have tasted pike. Far fewer, I'm sure, have ever tasted roast swan . . .

About ten years ago I knew a very keen angler called Tom Lang. He was a knowledgeable naturalist, excellent company and daft about fishing. From time to time he was a guest on my water and was rarely without trout or grayling even when other anglers were declaring conditions to be difficult or even impossible. ('Difficult' conditions to an angler are any time when he is not catching as many fish as he would like. 'Impossible' conditions indicate that he has not been able to catch anything.) As an appreciative gesture, no doubt, for his occasional evenings among the May flies or blue winged olives he invited me, along with several other anglers, to dinner at his home near Winchester. It was a Saturday evening in December and Rene, Tom's wife, was spending a fair amount of time in the kitchen before the meal whilst we drank mulled ale round a big roaring log fire. She was obviously going to a great deal of trouble to give us all a dinner to remember.

I noticed Tom and Rene exchanging knowing smiles when she came in to join us and when we moved into the dining room and sat at the glittering table Tom announced that Rene had an unusual treat for the main course, but we would have to wait and see. We were forced to eat our baked avocado and crab meat without further information and although no questions were asked by my fellow guests I imagined that, like me, they were wondering what could possibly be the cause of our hosts' obvious amusement.

When the main dish arrived we were a little puzzled at first. It looked like a large turkey decorated with a coronet of big white flight feathers. Tom had been right to promise an unusual experience for we were all about to join the Plantagenet kings in the delights of feasting on roast swan.

I had a vague feeling that, in those merry old days, the most unspeakable tortures were inflicted on any member of the peasantry unfortunate enough to be caught gnawing a swan's drumstick. They are still protected birds, of course, and as I joined in the appreciative cheers, I wondered what frightful penalties we might be called upon to face if Her Majesty the Queen ever found out.

Tom carved and generous helpings of the regal fowl were heaped on our plates. The roast potatoes, grilled tomatoes and mushrooms, petits pois, carrots Vichy and chipolata sausages which garnished the meat were quite delicious. The swan tasted like felt insoles that had been used to wrap fish.

Praise for the delicious repast arose from all quarters of the table and I was pleased to drink to Rene's culinary skills in copious draughts of white burgundy which were essential to ease the swan past the back of my throat. I was unable to contribute much to the table talk because it needed every scrap of concentration and willpower to get through the mountain of meat but I finally made it. I was so bemused by the effort that Tom had replenished my plate before I knew what was happening. I was well into the profiteroles with chocolate sauce before my eyes stopped watering.

The next day was Sunday and Tom came down to fish for grayling on the main river. As my wife and I were about to sit down to lunch we noticed Tom's car in the yard.

'Why not ask him to come in and join us?' said my wife. 'There's plenty of lamb and it's rather cold to eat outside today.' When I got to the yard Tom was sitting along the back seat of his car drinking tea from the cup of a steaming thermos flask. He declined my invitation with his usual courtesy then, peeling back the top slice of bread from a giant stack of sandwiches, he revealed a square of coarse dark brown material.

'I'll bet you and Rhona won't be eating swan for lunch,' he said with a self-satisfied grin; then, snuggling deep into his quilted fishing jacket, he sank his teeth into the first sandwich and took another swig of tea.

I learned on later enquiry that the swan had been acquired when, whilst Tom was fishing a Midland reservoir, he had seen the bird strike some overhead cables which killed it instantly. It had fallen almost at his feet. A sad little story but I could not help wishing that the stupid creature had kept its eyes open.

Chasing Rainbows

When we first came to this property it was part of a farm and, in order to avoid confusion particularly for the postal service, it was necessary to give it a new name.

The only bird I particularly noticed on my first visit was a heron, beating slowly up the valley above the plantation. The next time I came here I saw another one lift suddenly from a ditch and as it rose above the trees with slow, silent strokes of its grey wings, it was joined in the air by two more from the watermeadows. The name of Herons Mead seemed appropriate and since we have lived here our experience has done nothing to change our opinion on the subject. We have counted fifteen of these birds on one occasion on Night Paddock, the meadow just beyond the pool towards the river. My neighbour counted twenty-four near Manor Pool on the main river last summer.

These unusually large flocks, I feel sure, are family groups of parent and young birds, perhaps several families together but, although it is more usual to see single birds flying or fishing the river and side streams, there is no doubt that there are plenty of herons in the area.

Herons, as everyone knows, eat fish and it seems reasonable that if I want to enjoy the pleasure of living among these beautiful creatures I must be prepared to sacrifice a proportion of my own fish in return for the privilege. I have always accepted this situation and when stocking the water have allowed for certain depredations by these and other wild creatures. I have to confess that on several occasions, when I have found otherwise healthy trout – sometimes rainbows of two pounds weight and more – lying dead on the banks

[119]

with savage wounds, or seen similar fish in the shallows still feebly moving in spite of being transfixed from side to side, my convictions have wavered a little. I do wish that I could call a meeting and tell them to kill and eat my fish properly or leave them alone. At the time of writing, fish like those mentioned cost in the region of two pounds each from a fish farm and to find four or five such doomed specimens in one week – as I have done on more than one occasion – can be a bit upsetting.

It is senseless to become too sentimental about the endless chain of killings which takes place under natural conditions, for it is the way the world works and man's interference in the balanced ebb and flow of life so often produces unforeseen side effects which are much more horrible than nature's apparent indifference.

Sometimes I would like to suggest that the herons vary their diet with a few more water voles instead, for on occasion these are on their menu. It is not that I wish any harm to these gentle, inoffensive little animals but I am kept constantly on the run trying to repair the damage they do to the soft peaty banks of the pool and streams by burrowing continually into them until they collapse beneath my feet. They can be as appealing as rabbits and, like them, tend to multiply until they are a nuisance. Nature can take eons of time to balance herself but man does not live long enough to wait and that is the dilemma.

On many early mornings I have looked out of the bedroom window when pearly mists drift over the water and have not yet been burned up by the rising sun. I have watched the water and the banks for many minutes looking for signs of interesting or unusual activity and have then been surprised to see a heron rise at no great distance for, although they look such big birds when in the air, they can be a slender motionless line amongst the reeds when fishing.

Some mornings they are very obvious, standing on the Folly Island or by the alders on the rim of the pool and one morning I was surprised to see one fling itself from the bank and dive into the water like a gannet. It seemed to crumple when half submerged, like a bundle of thin sticks and feathers, but it emerged with a struggling trout, stalked to the bank and, throwing its head back, swallowed it head first. Its neck looked like a mis-shapen Christmas stocking as the bulky fish went down.

I banged on the window frame to drive it away and it lifted off across the water lilies and regurgitated the fish with a splash into the still water. The fish was floating when I went down but was quite dead. There was blood still on the gaping hole in its head. I almost wished that I had kept quiet. It seemed a pointless waste of good fish in every way.

Although herons can be expensive and annoying at times I am glad to have them around. They are nervous creatures and depart immediately at the slightest sign of human presence even at great distance. I have heard the story that they will die of fright if surprised at close quarters but, although some people seem to believe it, I feel it is like the assertion that a guinea pig's eyes will drop out if it is held up by the tail. The heron is so wary that he is not often so surprised – although one evening in late summer, returning from the river, I was almost frightened out of my wits when one panicked into the air from below the hump-backed bridge just as I turned on to it. He certainly did not die of fright but I almost did.

The Cormorant tree

Herons can be irritating but they do not make me angry. Cormorants do. Although they are sea birds, cormorants make quite long flights inland to find food in fresh water and, except at breeding time, they may be seen frequently flying up and down the river or perched like vultures on strategically placed trees near the water. They can be devastating on a trout river for they are strong underwater swimmers and voracious feeders and they are often present in fair numbers, particularly in the winter months.

Bob and I were walking along the edge of the pool one afternoon when we were surprised to see a cormorant's snake-like neck and head emerge from the water only a few yards away. We yelled and waved our arms in unison, dancing up and down at the water's edge. The clamour brought my wife running, white faced, from the cottage door but the cormorant seemed unimpressed. He dived again and came up on the far side of the pool. We ran howling round the bank, clapping our hands until they hurt and yelling blue murder. He merely swam away from us and went under again. We grabbed clods of earth from the bank and waited with soil dropping from both hands for the demon to reappear. When he did he was well out of range but we hurled our missiles anyway, shouting abuse. About forty ducks had fled the pool in panic. My wife was wondering if we were affected by some form of food poisoning.

A cormorant can swallow more than its own weight of fish in a day and we knew it but it was his defiance which needled us most.

The bird was under water.

'You go this way, I'll go that,' yelled Bob, grabbing more tussocks. We ran in opposite directions watching the water like Indian scouts waiting for a feather to appear. He popped up in the middle, between the Folly and Rose Islands. We hurled clods and abuse but they all fell short. We re-armed and sidled this way and that on opposite sides of the pool like wrestlers looking for an opening.

The cormorant rose about twenty feet from Bob and he hurled his turfs. To my amazement (and I think to his) his second missile caught the bird squarely with a dull thud and a ragged shower of water. He did not even change course before going under once more. Speechless now and with a sore throat, I was contemplating throwing off my clothes and settling the matter face to face when he surfaced once more, lifted heavily from the water and flew low across the surface before rising over the willows in an arc. We watched him in silent relief as he swung round over the church and behind the cottage roof. He completed a full circle and was coming in low and fast towards the gap in the plantation trees through which he had departed. He was coming back in and we could not believe it.

As he planed down I thrust two fingers into my mouth and gave an ear-splitting whistle which was all I had left. He missed the surface by a foot and rose over the high bank by the gate going like an arrow – a dark, strangely prehistoric little monster.

It is very difficult to keep either herons or cormorants away from open water unless one is prepared to spend endless time on patrol and for most people this is impossible. When trout are bred for the table or for the purpose of restocking streams and rivers they are kept in relatively narrow troughs of running water, called stews, and it is possible to net them over to prevent predatory birds from getting in. On a lake one can only hope for the best.

The menace of fish-loving birds seems to come in waves. Sometimes one does not see them for perhaps four or five weeks at a time then, for no apparent reason, they seem to be there every day until they are frightened off by some kind of drastic action. This is particularly so with cormorants and during one concentrated period it seemed that almost every morning when I looked out of the bedroom window I could see the serpent-like head cutting the water like a sinister periscope. If my wife was up first (which was usually the case) I would be shaken rudely into consciousness by a shout that a cormorant was fishing, or swallowing a fish or even calmly standing on the wooden seat by the water's edge, drying its wings.

One morning, when it seemed that the pool must soon be emptied by the impudent raiders, I awoke to see one of these strange looking birds standing again on the seat with his wings spread out, calmly drying them in the light breeze. Incensed beyond words, I crept downstairs for my gun. I was angry but tremulous at my own decision. I have noticed that cormorants invariably face the water when their wings are spread in this manner and it was only this fact that made it possible for me to open the window quietly without being seen. I took careful aim and fired. The bird simply fell forward onto the water, wings still outstretched. It did not even tremble. I was at the waterside in moments and feeling rather unhappy at what I had done. The bird was stretched on the surface, spreadeagled like a museum specimen. A trout, which later proved to weigh 2 lbs. 6 ozs., floated beside the bird. It had been regurgitated almost perfect. It was quite undamaged but was squeezed into a smooth tight oval package, like a convenience food in a supermarket.

I did not see a cormorant for more than a month after that incident, which made me suspect that one bird had been returning consistently, as I am sure is the case with our herons too. Later they were back again and have continued to be a nuisance from time to time. I have discovered that, like the swans, they are afraid of the

boat, however, and I have not been forced to use the gun since.

Last winter there was a layer of thin ice over half the lake surface for about a week and, not having seen any marauding cormorants for some time, I was surprised to be wakened with the news that another bird was on the rim of the water. They are the only web-footed birds in Britain which do not put out natural waterproofing oil to protect their plumage from the water. All members of the cormorant family find it necessary, therefore, to dry off their feathers after they have been fishing. It is a reasonable assumption that a bird which is standing with wings outspread has already eaten. Incensed once more but reluctant to shoot the bird, I opened the bedroom window and yelled at it. It jumped into the water, swam to the edge of the ice and jumped up onto it where some ducks and coot were also standing about. It then regurgitated a fish onto the ice and, like a heavily laden freight plane, it lumbered across the water and away into the air. The fish was quite still and showed not the slightest sign of life.

About an hour later, after breakfast, I went out for my morning walk and as I passed by the boathouse I wondered what sort of fish had been killed. I could not be sure at that distance so I got out the boat and rowed across. I was surprised and relieved to find that it was only a roach but one of considerable size. It looked about a pound in weight. I broke the ice with the bow of the boat until I could reach it. Throwing the fish into the bottom I returned to the bank and put the boat away. I decided to weigh it. As I was opening my studio door it leapt from my hand and flipped about in the grass. I could hardly believe my eyes after what had happened to it and the time it had spent out of water.

I hurried back to the hump-backed bridge where the flowing water was keeping the ice away and put the fish carefully into a strongish flow between two stones. With its head upstream and two small boulders preventing it from turning over on to its back I placed another behind its tail so that it would not be washed out by the current and there I left it. About five minutes later a slight movement of its gills was discernible and I went on my walk along the river.

When I returned half an hour later the fish had gone. It would require considerable energy to swim out against the strong flow, which was the only means of escape from its cradled position. It seemed impossible to believe but, as there was no sign of a floating roach anywhere about the lake, I was forced to conclude that it had recovered and returned to the deeper water.

Fishermen of England

Since I have been fortunate enough to have some water of my own
I have not fished nearly so often as I did when I paid to fish other
people's lakes or rivers. This is not due simply to the perversity of
human nature but to the fact that my interests have gradually
changed from the desire to catch fish to the even more intense
pleasure of preserving them as part of the life of the river as a
whole. If fish are needed for the table I still enjoy angling for them
as much as ever, perhaps more so, but watching a big trout in the
clear water below the silver birches or another rising to fly beneath
the blue bridge gives me far more pleasure than I could get from
catching them.

There is no doubt that for most anglers the pleasure of being
near to water in beautiful, peaceful surroundings is half the
attraction of the sport and for some it is the most important feature
by far. I once knew an angler who rarely, if ever, threw a line upon
the water. Charles Bicknell was what is often referred to as a
successful businessman. He was past middle age, of portly build
and spent one day each week fishing on the Test. He would arrive
for his day's relaxation in a Rolls Royce Silver Ghost driven by a
smart, liveried chauffeur named Parker. The routine was always the
same for I often met them on my morning walks. Charles would
stretch his legs after the journey by walking the thirty paces from
his car to one of the rather tatty but comfortable armchairs in the
club-house. He would lower himself into it, take a long swig from
an expensive silver flask and go to sleep.

Outside on the river bank Parker would unload the fishing tackle and set about assembling it. He was a tall, spare, likeable man of about thirty and always seemed anxious to have someone to talk to. He made no secret of the fact that he thought anyone who chose to spend their leisure hours in such a god-forsaken, out of the way place must be a nut. I would come across him sitting in the grass, fumbling with the expensive tackle like a five-year-old trying to knit. He would give me a rueful grin and look heavenwards as the long rasping snores of his master floated from the club-house door.

One morning I found him sitting on the bank amid a tangle of tackle that looked like a dilapidated heron's nest. He was staring up the quiet river with an expression of utter boredom. I had never seen anyone with fishing tackle in such a mess before the day's fishing had even started and I offered to help him sort it out.

'No thanks,' he said. 'I've got all flaming day. When that lot's put together there'll be damn-all to do.' I longed to see him cast a fly but I was never able to stay long enough to see Charles emerge for his weekly exercise.

Later in the summer I was walking down the river bank in mid-afternoon when I saw Parker some distance ahead lashing at the water with a rod as if he was trying to beat something to death with a whip. He stopped when he saw me approaching and gave me a resigned grimace.

'I don't mind putting his flaming tackle together,' he said, 'but to hell with doing his fishing for him.' He threw down the rod, sat down hard on a nearby seat, folded his arms and stared fixedly into the distance. As I passed the club-house I looked through the open door. I could see Charles in an armchair with his head back and his mouth half open. His snores floated gently out across the golden river.

Charles Bicknell would certainly qualify as one of the most relaxed fishermen of my acquaintance but there were others; like old Jack Grey, the only man I have ever known to carry a folding garden chair with him when he was fly fishing. He was old enough, likeable enough and expert enough to walk up the river banks of summer clad only in a pair of long khaki shorts, and produce nothing more hostile than amused smiles from his fellow anglers in their full dress tweeds. He had reached the venerable stage of life when a man dozes off easily, particularly on a flower scented, droning summer afternoon, and he liked to be close to his beloved water. He was so close on one occasion that when he nodded off he pitched forward head first into the river. It did not ruffle him one jot. He simply smiled and pointed out that wearing only a pair of shorts had many advantages.

One August afternoon I was fishing along the carrier with my brother, Alan. The sun was bright and hot and we were walking up the bank between high reeds and serried ranks of waterside flowers. The perfume and colour were as exotic as any cultivated garden could produce and bees hummed amongst the massed pink, purple and yellow blossoms. The air was alive with butterflies and nebulae of delicate ephemeral insects danced away their brief existence on gossamer wings about our heads.

I was watching the water without much hope of seeing any rising trout or grayling in the bright conditions when Alan stopped so suddenly that I crashed into him.

'What a stupid place to leave a rod,' he said, pointing to the narrow path through the foliage a few paces ahead. A delicate split-cane fly rod was protruding from the tall reeds to our left and lying across the ground where it must inevitably be trodden underfoot by anyone walking along the bank. It was fortunate that he had spotted it in time. He picked it up and we were speculating on what strange circumstances could induce such carelessness when I noticed a pair of rubber wading boots sticking up at a steep angle from the reeds ahead.

We hurried forward and were shaken to find a body lying on its back deep in the reed bed. The ground sloped down to a drainage ditch there and the head was almost two feet lower than the felt-soled boots and only inches from the stagnant ditch water.

'It must have been his heart or something,' said Alan, whitefaced. 'I wonder if he's still alive?'

We pushed through the reeds and looked at his face. His eyes were closed and his mouth was half open. It was Captain Parton who had been a member of the club for several seasons but I had not immediately recognised him from such an unusual angle. His clothing retained its customary neatness and his tweed deerstalker was still centred but gravity had raised it a few inches from his balding head. He looked as if he had been pole-axed from the front and fallen backwards down the slope like a doomed tree.

My medical knowledge in circumstances of stumbling across a suspected corpse is confined to loosening its tie if it is the right sex and, in either case, checking that false teeth, if any, are not wedged halfway down the windpipe. I undid his tie and opened his collar. Angler or not, I did not fancy fishing down his throat so I was relieved to see a neat row of sparkling incisors in what appeared to be the right place below his moustache.

Even in this lowly state he retained a certain military ferocity of countenance and I felt uneasy fumbling around his jacket to find his heart.

'He's still alive, I think,' I said, turning to my brother. 'Let's get him onto the path and I'll run for help.'

Alan took hold of his waders but the black water-filled ditch made it impossible to get behind his shoulders so we settled for one foot each. We pulled steadily and he slid with surprising ease up the thick, springy mattress of reeds and on to the grassy path. As we turned him onto the flat ground by hauling more strongly on his left leg his eyes opened and he said, 'What the hell do you think you're doing?'

We dropped his legs as if we'd both been shot by one bullet. He sat up unsteadily, blinking little blood-shot eyes against the fierce sunlight.

'Are you all right, Captain Parton?' I enquired in a state of shock.

'Why the devil shouldn't I be all right?' he snapped.

'Oh, we were just passing,' I said, 'and saw you . . . er . . . saw you . . . er . . . lying down and we wondered if you were all right.'

Away from the rank pungency of the ditch water the potent fumes of whisky billowed upwards in the corridor of foliage, overpowering even the heady perfume of the chalk stream flowers.

'Get no fish in this light,' he muttered. 'Might as well have a nap.' His eyes closed and he fell backwards with his head in a dense patch of water forget-me-not. We laid his rod by his side and walked on, stiff with shock and embarrassment. The ambiance of the afternoon had vanished and we crossed the first bridge and walked back down the opposite bank so that we would not have to step over the captain's prostrate form.

Although he fished with us for several more seasons, Captain Parton never referred to the incident and I often wondered whether he remembered what had happened. If not, he must have been puzzled when he finally recovered to find his shirt and tie undone and his hat on a mass of flattened reeds down by the ditch. I comforted myself with the thought that things could have been very much worse. At least I hadn't removed his teeth.

There is, of course, another school of anglers who approach their sport with a very different attitude. They tend to be breathing heavily before they have got their tackle put together. They rush about expending more physical and nervous energy than would be shed in a full day's work and when a rising fish is located, discover that they have by-passed several rod rings when threading their line and have to take it all apart and start again.

Angling may well be the contemplative man's recreation but it can also stir up strong passions. One summer evening along the river I was introduced to three young gentlemen who were guests of a fellow member of the club. They were immaculately dressed in lounge suits, for they had never fished before and they had borrowed tackle for the occasion. They were utterly charming and could scarcely find words to express their delight with the beauty of the river, the scenery, the weather and the warm friendliness of the club members.

There was at that time a phenomenon on that stretch of water which was known as 'the boil'. It occurred at almost exactly nine o'clock each evening when every fish in the river would decide to rise at the same time. It was the most extraordinary sight. I have never seen anything quite like it before or since but we had quickly discovered that during the crazy half hour which followed it was impossible to catch any of the fish. Every known pattern of fly had been used and many monstrosities which were nameless, but without success. Theories about what the fish were eating were as numerous as the flies on the water, but they produced no results. It was impossible not to react to the challenge of the boil but we all knew more or less what to expect.

The three young gentlemen apparently did not. It was the most beautiful balmy evening and the bells of the distant abbey drifted across the sunlit water meadows. The trio were fishing rather close together about 200 yards away when the boil began. There were audible shouts of amazement and joy. In a matter of minutes the shouts became noticeably louder and somewhat different in tone. Words which did not quite fit the idyllic pastoral scene became clearly audible. I noticed several other members of the club abandon casting to watch developments.

Within the next half hour I heard every colourful expletive that I had been introduced to in five years of army life and a few more modern ones which were new to me. I could see wide grins on the faces of other fishermen and before the show ended one or two were dabbing their eyes and leaning against trees for support. I was walking back home later when one of the club members caught me up.

'They can say what they like about young people today,' he said, 'but they were three likeable lads.'

There tends to be a fairly high proportion of military men amongst the game fishing fraternity and, having spent a considerable part of my own war service as a private soldier in an infantry regiment, I have to own to laying the responsibility for many long hours of cleaning greasy cookhouse utensils and other miserable experiences at their doors. A little to my surprise, however, I have found them, for the most part, to be friendly companionable men.

Major Hebden certainly was. He introduced me one day to a quiet unassuming man whom he had brought along as his guest. His friend did not readily fit into my mental picture of a high-ranking officer and I was surprised to learn that he was a general.

His face was badly scarred by some dreadful physical injury which had clearly not affected his good nature. The major told me later that the general had been the sole survivor of a plane that had been shot down off the North African coast in 1943. He had fallen from the aircraft before it crashed into the sea but, having no parachute, he had been almost torn apart when he struck the water. By some strange quirk of fortune he had been picked up whilst still alive and sent to hospital where he had suffered long years of repeated operations before being restored to his present condition.

'Do you know,' said the major, 'I have known Arthur for more than forty years and have never heard him complain once. He is the bravest man I have ever met.' It was one of those touching human stories that makes one feel humble and ashamed of thinking unkindly of any group of people whom one does not really know.

One evening, about a week later, I decided to go down to a stretch of the river where I had seen some good rises under the far bank when taking an evening walk. We did not stick to fishing individual beats in those days but shared the whole river on a gentlemanly basis. I was out of luck for, as I approached the river, I saw an angler dashing madly down the bank with his rod arching strongly towards the water. There are a number of scattered alder bushes along this stretch which are a menace to anyone playing a fish if it should decide to run strongly up- or down-stream and the running figure was fast approaching one of these obstacles.

It was the general and he was in a sorry state. His face was red and glistening with sweat, his hat was gone and he held his rod high in the air with both hands as he ran. I was almost trampled as he rushed by, wide-eyed with terror. All his efforts could not raise the taut line above the highest branch of the alder and it snagged for a terrible moment before being pulled free. The branch sprang back into place and the unseen fish, with the general attached, continued downstream.

In the open stretch before the next alder there was a check and the general began to reel in. I hurried down the bank to see if I could help.

'It's . . . the . . . biggest . . . trout . . . I've . . . ever . . . seen,' he gasped. 'I've had him on for about twenty minutes. I *can't* lose him now.' He looked all in.

His reel gave a little screech and line cut the water back in the direction he had come. He dashed after it trying to keep in touch – straight for the alder which was waiting like the angel of death. Even from the back I could see that the man was beginning to crack: his co-ordination was going and line whistled deep into the branches. I could hardly bear to look. I ran back to him, more as a gesture than with any hope of helping, for line was still peeling off the reel into the tangle of branches. I gazed up the river, astonished at the power of this giant trout which I had not yet glimpsed.

Suddenly it leapt from the water in an arc of silver and fell back with a loud slap. The line went slack on the reel. I was sure I had seen a salmon leap but I was not disposed to argue at such a moment.

There are times when human beings can find no words and the line was recovered in silence. Nothing would have induced me to look at the man's face for it would have been a tasteless intrusion into private grief. Glad of the excuse, I recovered his hat and landing net from further along the bank and returned them to him. I wanted to offer a word of sympathy but could not.

Finally, he turned briefly towards me and said 'Thank you. Goodnight,' and walked away down the river looking old and tired. If I hadn't known better, I could have sworn that he was crying.

Chicken Feed   Although the ducks and other waterfowl are the most numerous inhabitants of our establishment, they are by no means the most powerful group. As soon as they are foolish enough to step ashore they are subject to the despotic rule of the bantams, led by Henry II (or Harry Hotspur, as he is more often called).

He has ruled the roost and the territory around it for nearly nine years and advancing age has done nothing to undermine his self-opinionated attitude or cool his uncertain temper. Currently, he has six little wives as plump and pretty as partridges and three sons who already show signs of inheriting their father's bellicose character. Like Joseph, he wears an exotic, technicolor dreamcoat and is convinced that he has been chosen for great things

They live in an outbuilding in the yard and never stray more than a hundred yards in any direction. His domain is small but Henry's rule is absolute. His father, Henry the first, was equally courageous and ever ready to teach any interloper a quick lesson or to lay down his life to protect his hens. Unfortunately, he laid it down permanently when a thoughtless neighbour drove into the yard and released two labradors from the back of his Landrover.

The most amateur bird mimic can bring our present hero bursting from the pop-hole like a spangled cannonball, dancing and weaving on tiptoe with wings held low and ruff spread stiffly out round his head. He looks like an Elizabethan gentleman in court dress warming up for a duel. His spurs are so long that he frequently trips over them when he runs.

One afternoon I heard a commotion in the yard and saw the hens running madly hither and thither, flapping and squawking in panic. In the middle of the yard was what at first sight looked like an Edwardian lady's hat – a low mound of tatty but brilliant feathers rolling this way and that and changing colour in the sunshine. Little clouds of dust and feathers exploded into the air.

Slightly less than half the 'hat' turned out to be Henry who was lying on his side holding grimly on to an even more extravagantly feathered cock that I had never seen before. Having no previous experience of breaking up cock fights, I ran up to them waving my arms about and shouting 'Break it up lads!' The newcomer, who had a vice-like grip on Henry's neck, let go with a strangled squawk and fled under the conifer hedge.

Henry staggered drunkenly to his feet, turned round in two or three unsteady circles and then, twisting his neck like a piece of distorted driftwood, delivered an earsplitting crow of triumph. He stumbled away with half his neck naked and the yard was left deserted except for pathetic little bundles of bright feathers that blew this way and that in the breeze. I still have a few trout flies made from the hackles plundered from the battlefield.

Later that evening a lady from a nearby village called us on the telephone.

'Did you find Arnold?' she enquired.

'We haven't lost anyone called Arnold,' I replied, puzzled.

'My bantam cock,' she explained. 'I'm moving house and had to find a new home for him. I just can't bear to have him killed and I knew you kept bantams, so I dropped him in your yard this afternoon. It's such a relief to know he will have a good home and some little friends to play with.'

I couldn't find words for a moment or two but I suddenly blurted out, 'Oh thanks very much, Mrs Potts. Yes, he's out there right now and I'm sure he's going to be all right.'

Out there he certainly was, peering from the depths of the cupressus trees whilst Henry patrolled the yard to make sure he came no nearer. It was clear that Henry was not prepared to welcome rivals for his hens' affections and I was worried that Arnold might starve or be taken by some predator if he was left out to fend for himself.

I tried every trick I could think of to trap the new cockbird whilst my own flock watched the proceedings from a safe distance, but without success. I scuffed my shoes, stained the knees of my trousers and ruined a perfectly good jacket on barbed wire before Arnold, three days later, was safely in a cardboard shoebox. Douglas gave him a great welcome in his orchard over at Braishfield and he quickly became patriarch of a thriving new tribe.

It is understandably an emotional time for any fowl when they hear the rattle of the corn bucket and I was not greatly surprised, at least not after the first few days, when they arrived on the scene from every direction – including one o'clock high – as soon as the lid of the corn bin creaked. I was, however, rather taken aback when Henry, looking like a pigeon in drag, took to nipping my heels to speed up the service. It was a blatant case of biting the legs that fed him and it made me jumpy.

After a week or two of practice he was obviously enjoying the routine and I began to feel that if he were not checked he would one morning have me down. The day inevitably came when it was time to call his bluff. He squared up to me, or, rather, to my Wellington boots, as I was crossing the yard. I stopped and turned towards him. In a moment his wings went low almost brushing the gravel, his ruff rose round his head like a halo and he started dancing this way and that on his toes. He bounced round and round me, bobbing and weaving like a pugilist and every few seconds he launched the full weight of his attack (about twelve ounces at a guess) at my feet.

It is difficult to know how to fight back in these circumstances and he was getting me rattled. In a moment of crass stupidity I held my arms stiffly down and backwards like his wings and started dancing round *him*. It was a fatal mistake. He threw himself with renewed energy into the fray whilst the hens looked on like a crowd outside a pub watching a brawl. I felt I was losing dignity as well as ground, for I was flapping my arms by now and hurling abuse at him. He was a ball of feathered fury. Obviously he was taking the whole thing far too seriously and the first feelings of panic began to wash over me. I realised with horror that he regarded it as a fight to the death. I turned and ran with Henry scattering the gravel behind me. It was a relief to find that he pursued me only as far as the entrance to the yard then, jumping onto the wall, he went through his clarion call of honour. I couldn't hear the hens cheering but they had the right.

I will not deny that I was apprehensive the next morning. The rain was pouring down and looking through the glass of the back porch door I could see the chickens huddled beneath the dripping eaves of the stable waiting for breakfast. I picked up the black

umbrella from the corner of the porch and stepped out. The pathetic little group burst into life and charged across the yard like football hooligans. They were about ten feet away when I opened the umbrella. The effect was magical.

They flew screaming in all directions. Henry went straight over the garage roof and the yard was empty. Protected from both fowl and flood by the umbrella I scattered corn in safety and returned to the house unmolested. When I closed the umbrella at the porch door they all reappeared like conjurers' props and fell on the grain as if nothing unusual had happened.

For some time after that I carried the umbrella over my head at feeding time until one bright sunny morning I noticed the postman giving me an odd sort of look from behind the lilac tree. A bit alarmed that strange stories might be spreading round the village, I abandoned the practice and was relieved to find that Henry continued to treat me with reasonable respect. If the situation should ever deteriorate again, however, I know exactly where to lay my hands on the umbrella!

Animals are fascinating and most of them are very beautiful indeed but, with the exception of certain domesticated creatures, I much prefer them to remain wild and retain their natural dignity and independence.

I know that many people take great pride and experience great joy in taming wild animals but it is not for me. I do not much care for circuses for this reason and, although I am fully aware of the pleasure that zoos provide for millions of people and of the great work that zoological societies do to preserve and even reinstate threatened species, I am not drawn to the idea of animals being dependent entirely upon human beings for their continued existence. It is a paradoxical situation, for I have spent many hours drawing animals which I would never have been privileged even to see if it had not been for the fact that they were made available to me in zoos and I can think of few more worthwhile objects in life than saving animals and birds from extinction and releasing them once again into their natural surroundings. Yet I still find the sight of a magnificent Siberian tiger in a cage sad beyond words.

It is not that I cannot be influenced by logical arguments, it is simply that I am always swayed back the other way by the reality. This is why I do not pinion my ducks and, consequently, why I lose so many. It is why I have taken the wire mesh from the entrance and exit of my pool and do not now have such big trout. It is also why I do not even feed my ducks or fish any more.

The valley is full of wild duck that thrive on the natural food of the river and ditches and meadows and I think my own ducks will enjoy better lives doing the same. When I first introduced them it was necessary to feed them daily in order to establish the pool as their base but, as with so many good intentions, this practice soon produced undesirable effects. I have described elsewhere the problems which arose from feeding trout pellets to the fish and how the ducks became suicidal in their efforts to steal this food, but even this rich, extra food supply was not enough to keep them happy. They soon learned that the feed bin which contained their grain and that of the bantams was in an outhouse in the yard and they began to spend less and less time on the water and more and more time on the scrounge.

I first became aware of the problem when their heads began to turn black. I was alarmed and puzzled, for it was like some awful disease and it spread rapidly onto their necks and backs. I managed to catch a white duck at feeding time and found that it was filthy black sump oil as thick and viscous as butter. Once I knew what they had on their heads, it took no great powers of detection to work out where it came from. Cars were always parked in the yard

rather than at the front of the cottage and the ducks now preferred to meander in and out and under cars rather than reeds and bridges, for the yard was the fountainhead of all good things and they did not intend to go far away from it.

I tried every trick to discourage them. I chased them back to the lake at half-hourly intervals. I kept the gate closed (they had become reluctant even to fly over it if they could avoid the effort) and even banged and bounced tin-cans on the stones to scare them away but it did no good. So I stopped feeding them. They hung about miserably for a day or two and then realised they would have to get back to work – the boom days were over.

Since then they have lived where they belong and their heads are the right colour again. They do not panic when approached as truly wild duck would do, but they are more alert and wary – independent of human hand-outs – more beautiful and, as a bonus for us, the yard is much cleaner. When the bantams sit in a row, peering through the French windows, it is very hard to resist throwing them some scraps but we have learned our lesson with them too. Fed a limited amount of grain once a day, they spend nine tenths of their time scratching about for more food and keeping the garden free of weed seeds and pests and getting fresh air and healthy exercise into the bargain. A few scraps through the cottage door will reverse their timetable instantly. One tenth of their working day will be spent hunting and nine tenths peering reproachfully through the windows. They are cute in more ways than one and as cunning as a gang of medieval beggars. They have outwitted us so often in extra-ration-games that, even when Henry trips over his own spurs and falls on his face, we suspect it is a ruse to drum up grains of sympathy.

The problem of looking after animals was nicely summed up by an old lady who used to be a close neighbour. She was having a cup of coffee and a chat with my wife one morning and I tried to explain away the row of beady little eyes watching us through the sitting room window. She quite understood, she said; it was the trouble with the world nowadays. Everyone wanted more wages for less work and she didn't know *where* it was all going to end.

The Walnut Tree   If I look up from my desk I can see the water and the valley beyond framed by the lower branches of the walnut tree. Its leaves reach out like seven-fingered hands to within a few feet of the studio window and its higher branches rustle and sway above the red tiles of the roof. Thirty feet from the doorway the silver-grey, deeply rutted trunk, which measures ten feet around its girth, rises only twelve feet above ground-level before spreading out in a fountain of great limbs. They writhe and twist sixty feet into the air and spread nearly seventy feet from side to side. It is magnificent. It makes me eternally grateful to those of our forebears who planned for future generations rather than for their own. It is certain that whoever planted the seed and tended the sapling did not live to see it reach maturity, but I have a feeling that he knew exactly what he was doing and enjoyed great satisfaction in doing it.

When I look at the tree in the dark days of winter, its huge green-black skeleton silhouetted against the ashen sky, or hear its tracery seething in a westerly gale as I lie snug and warm in bed, I wonder who it was that planted this giant for so many generations to enjoy. And in the balmy days of summer when its leaves are overlaid like the breast feathers of a great bird to form high domes of rounded foliage, I wish I could call back this gentle spirit of the past and say: 'This is your tree. Look at it now, for it is gracious beyond words.'

Perhaps he lies among the lichen crusted headstones in the little churchyard across the water. The red-tiled bell tower will have seen him plant this tree and watched it grow, through years of human turmoil and restless change, to quiet perfection. A tree expert from a famous arboretum nearby has estimated its age at about one hundred and fifty years.

It is possible that the mighty, close-grained limbs sprang from a nut dropped by accident or buried by a squirrel, as they still are every autumn, but I doubt it. Its young shoots would have been eaten off by cattle or deer or rabbits before they could develop into a maturing tree, unless they were protected by human effort.

One Sunday morning a stranger knocked at the cottage door and asked shyly if he might look around the place. He told us that he had lived here when a child and had not seen his old home for many years. He was delighted by every detail that was still unchanged: a post in the hawthorn hedge that had held the wooden gate he once played on; an apple tree stump in the rock garden that sent him climbing again in memory along hoary branches; but most of all he loved to see the walnut tree again. It was, he said, exactly

as it had been that day in 1904 when he first saw it as a child. He had lived beside it for nearly twenty years and recalled each great limb with delight. He showed us where his mother's washing line had hung from bole to branch-end and where he sat in summer shade against the trunk.

Returning to the scenes of childhood is so often a disappointment that it was a great joy to us to see his obvious approval of what we had done and his happiness at what we had not changed. I had the feeling, however, that if all else had succumbed to mindless 'progress' he would have thought his journey worthwhile to stand beneath the walnut tree again.

There can be no doubt that the tree has grown considerably in the past seventy-three years, but the passing of time adds grandeur to our memories of childhood also and this probably explains why the tree was all our visitor had expected it to be.

I must admit that when the gales roared in across the valley and bent the high poplars like fishing rods, I felt some unease sitting here beneath the great twisting limbs, for I knew that if one had fallen it could not have missed this building, so I decided to contribute a little to the welfare of the tree and a great deal to my own confidence at the same time. The deeply grooved bark had split in several places where the branches forked from the bole and there was a little rotten wood in these dark clefts. I consulted a company of tree surgeons and found to my delight that not only was the tree completely sound but that the men who came to see it considered it the finest walnut they had ever examined. The cracks were cleaned out like small fillings in a tooth but they were not filled in for they explained that fillings in trees work loose in time and the rain seeps down behind to rot the wood unseen. They painted the shallow hollows with a water repellent so that rain now runs straight out again and the cambium layer can reseal itself over sound timber. As an extra precaution to strengthen the heavy branches in rough weather and protect my studio at the same time, they connected the main limbs about halfway up their length with three fine steel hawsers. They are so cunningly installed that they are difficult to detect even by those who know what to look for.

Compared to many trees the walnut is rigid and unyielding in high winds and it sheds its outer twigs freely in wintertime. Perhaps this characteristic has given rise to the old country rhyme:

> A woman, a dog and a walnut tree
> The more you beat them, the better they be.

It seems reasonable that the tree at least might be better for a good shake out, but he would be a giant indeed who could beat more than the lower branches of this tree.

As a town child I had always associated walnuts with Christmas for it was the only time we ever saw them. They were synonymous with the rich smell of tangerines and cigars, roast pork crackling and Christmas pudding. I did not eat many for they were tedious to extract from the shells and I was more interested in the shells than the fruit. I begged each member of the family to be careful how they cracked them open and to give me their shells undamaged. Each nut produced two little boats if treated with care and by Boxing Day I would have a fleet of these little craft with matchstick masts and paper sails. Both halves of the shell together made a perfect container for treasures such as coloured glass jewels or a tiny celluloid doll from a cracker laid in cotton wool. I had no shame in playing with dolls, provided that they were small, because they became elves and fairies and were vaguely connected in my mind with fairy tales and magic. I remember sealing up a jelly baby in this way and forgetting it for some time. When I opened it up later it was a revolting, sticky mess. As the New Year advanced, the boats and treasure chests would gradually disappear; they would be lost, broken or forgotten like the conkers of the previous autumn.

In the autumn now the lawn beneath the tree is cobbled with hundreds of nuts and there is a rich harvest of fruit for all comers. There are fat and lean years of course, but we have never had a season which did not bring the rooks and jackdaws and squirrels to join the ducks and bantam chickens in an orgy of looting.

They arrive at first light and work until evening carrying away the booty. The birds bicker and squabble in the high dome of fading leaves; the squirrels run up and down the bole and branches collecting their winter stores and burying them almost anywhere. They inter them all over the lawn, in the rockery, in the flower tubs, under the dry wall and deep in the tangled undergrowth of the plantation.

I suspect that the bantams do not get much nutrition in return for all the energy they expend in strutting about and bullying the interlopers on what they consider to be their private territory but they do not let all the nuts go to the Johnny-come-latelys. I had never imagined that ducks were equipped to cope with walnuts even if they wanted to, but I was wrong. They do not spend much time picking, pecking or gnawing like their rivals, nor do they bother to carry them away. They simply rummage in the grass and leaves until they come across a nut and swallow it whole.

When I first realised what they were doing I was convinced that I would soon be collecting ducks ready-killed and pre-stuffed with walnuts but I was wrong again. Their crazy behaviour seems to do them no harm even when their breasts feel like a bag of ping-pong balls and their faces are stained with juice like badly made-up clowns.

Getting the bulky shells down their throats is not achieved without considerable effort and at times I can hardly bear to watch the necessary contortions. The mind boggles at the power of the digestive juices which must be available in a duck's gizzard. On one occasion I was so uneasy about their nut-swallowing that I took a bucket of tail corn down to the steps and made great play of broadcasting it. They seemed reluctant to leave the tree but at length they waddled down the slope to the water's edge and pitched into the grain. One white call duck, however, was obviously in dire straits. She was unable to close her bill because of a walnut which was jammed halfway down her throat and she seemed unable to swallow or reject it. She stood there desperately jerking her pretty little head back and forth, trying to clear the monstrous obstruction.

I felt sure the end could not be far off if I didn't do something quickly. I picked her up and found I could quite easily grip the nut between forefinger and thumb and I pulled as hard as I could. Her neck looked as if it were being stretched on the rack but the nut did not move. I had visions of driving madly down to the vet with her, howling for help like a fond mother with her infant's head jammed in a saucepan, but I felt sure she would not survive the journey even with a police escort. She struggled desperately to get out of my arms – the death throes, no doubt – and I knew I had to make a life-or-death decision. I put my finger on the nut and pushed. It shot down her throat like an oyster and joined the others already grinding about beneath the feathers. With a final effort she broke free and fell to the ground, struggled to her feet and, flinging herself into the feeding multitude, she started shovelling down the grain as if she had not eaten for a week.

They cleared the corn in minutes and almost ran back to the tree to pack in some more walnuts. Henry, the bantam cock, scattered a few feathers off their backs with his fearsome spurs just to remind them of his superiority, but they hardly bothered to look up. It was a cross they were often called upon to bear and the nuts, it seemed, were worth it.

It is at this time of the year that the squirrels are most in evidence and I am a bit surprised to find that my feelings about them are rather mixed. They are all grey squirrels, of course, and whilst they are visually very attractive as they scamper along the branches and leap from tree to tree setting the long twigs swaying, they become so bold and aggressive when the nuts are about that the less flattering name of 'tree rat' seems more justified. Sometimes they will be gone like a will-o'-the-wisp when approached but are just as likely, in autumn, to sit up and call vituperative obscenities at anyone who comes too near. They will look down from the upper branches and hurl chattered curses as if they were beside themselves with hatred.

One warm Saturday as we were sitting round the table eating the mid-day meal, my wife caught everyone's eye by putting a finger to her lips, raising her eyebrows and then pointing downward below the table. We all went silent, waiting to see we knew not what. A grey furry creature ran out from under a chair, round the kitchen and back under the table again. It was a grey squirrel which must have come in through the open French window. It seemed quite unafraid of us and was soon taking food like a mischievous monkey. It leapt on to my daughter's shoulder and rummaged in her hair. Its bold familiarity, without formal introduction, was startling and slightly sinister for it was not gentle in its movements but tore about the house and leapt upon us with wild abandon, keeping up an aggressive whittering chatter. It pillaged the place for perhaps twenty minutes before dashing out on to the lawn and disappearing up the walnut tree amongst the thick leaves. We shut the door and took time to catch our breath. A few minutes later we were astonished to hear a vigorous knocking and scratching at the glass and to see him peering angrily in as if he had been shut out of his own property.

Woodpeckers, both greater spotted and green, are often up in the high branches searching the crevices for insects, although they prefer the old half-dead apple tree on the lawn where there is more rotten wood. They can often be heard too, drumming on the dead, ivy-clad alders like bursts of distant machine gun fire.

I have never been quite sure why the green woodpecker, with his sour-apple body and ripe crimson head is called the 'rainbird', but I am glad that country people through the ages have dreamed up such attractive names for the flora and fauna of Britain. 'Yaffle', his other country name, has exactly the right ring to it, for his laughing call can often be heard as he loops across the water from the woods beyond the church and swings up into the branches.

Sometimes in winter I glimpse a treecreeper on the walnut tree, working his way up the deep ravines of the bark with quick movements like a little brown mouse, or a blue nuthatch, like a tiny kingfisher, working his way down.

The magpies do not come often to the walnut for, like all their kind, they are wary and highly intelligent. They usually hang about in the tall trees of the plantation, watching everything below with eyes like an eagle. If they spot any booty they wait patiently for the right moment and make a sudden raid, grab all they can and are back in the fastness of the high poplars in seconds.

I have never seen a magpie in the walnut tree without him seeing me first so that he is already on his way to safety by the time he registers on my eye. When this happens I know that some devilry is afoot and search around for bantam eggs, for I know that he has seen something close to the house that he can turn to advantage. I usually find some too, for they are dropped anywhere by our little hens at any time; on the lawn, in the gravel of the yard, in the wheelbarrow or in an old cardboard box on the lawnmower, and once behind an armchair in the sitting room. Many bantam eggs have been rescued when magpies have been spotted lingering with intent but the black and white devils usually get there first. They have more time than the rest of us to watch for eggs to pinch and their reflexes are a hundred times faster.

Last spring a pair of jackdaws nested in the chimney of the central heating boiler and almost killed the family with poisonous fumes although they themselves seemed unaffected by either heat or toxic gases. I tried various ways of getting rid of them without success and finally I was forced to get out my gun. After much pitting of our wits against each other (the birds knew as well as I did what was going on) I managed to shoot one of them with a single bullet from behind drawn curtains of a bedroom window. It flew straight over the stable roof and dropped like a stone in the paddock beyond. I rushed down the stairs and across the yard to the little field gate. Not fifteen seconds had passed since I saw him fall, but a magpie was already standing on his prostrate form and pecking at him.

The walnut tree stands high above the cottage and outbuildings and looks down upon the water. The lake and the tree have much in common for they each provide food and shelter for a wide variety of birds, animals and insects throughout the seasons. If no misguided opportunist destroys it for the value of its timber, it will stand here for many years to come, a symbol of continuity and stability in a shifting, uncertain world – a delight to the eye and a vital part of the delicate mechanism of life.

Last spring my wife found a little walnut tree growing in the flower tub by the back porch door. The nut must have been buried by a squirrel and forgotten. It is nearly twelve inches high now, a healthy, thriving little plant. When it is big enough we will transplant it to a more suitable site and take care of it. This seedling will not reach maturity within our lifetime but it seems the least we can do to show our appreciation for the pleasure that our great tree has given us.

End of a Decade

It is December, mid-afternoon and the light already fading. Outside the world is steely blue and grey and a bitter wind is moaning in the bare branches of the plantation. The autumn leaves lie like a dark, discarded skirt around the walnut tree and lift fitfully into the wild air, bowling and flying down the slope of sodden grass into the dark pool. Three ducks, ghostly white against the dark blot of Rose Island, and a raft of toy-like tufted, bob and bounce on the crinkled water like a flotilla of boats on a stormy sea.

A sudden burst of icy rain sprays the window like lead shot and dead twigs rattle on the roof tiles. My thoughts turn to the evening meal by a roaring log fire and the flickering light on curtains drawn against the winter night. I feel a pang of pity for the solitary wood pigeon swaying on a comfortless perch in the high poplars.

Deep down the primeval fear of the winter solstice lingers still and touches me with its cold fingers. Light and warmth have fled before the north wind and left the green and golden world in a grip of ice blue. There is little outward sign of fish in the pool now and the multitudes of the warm summer nights have gone to ground. It will soon be Christmas.

Ten years have passed since we cut the wild hedges and carried the first stones on Christmas Day so long ago. Thousands of young creatures have been born in or near this stretch of water and most have lived their brief spell and moved on or died here, to make room for those to come. Harry Hotspur is dead, taken by a predator in the field behind the chickenhouse. His bright feathers are scattered and his hens have passed to one of his sons. Our own children have grown up and are independent.

We are in the trough of the year but there is no end and no beginning, for tomorrow may be bright with clean winter sunshine and there are tiny wild flowers still, crouching in the cold grass. The owl is still hunting in the night and when one is roused in a warm bed by the unearthly scream of a vixen in a distant wood, it is the promise of warm new life to come.

The warblers' nests are ragged in the reed beds. The cuckoos and the swallows are long gone but the wrens are snug in their old nests beneath the boathouse rafters and the heron still stalks the crackling cat-ice in the ditches of Night Paddock.

The tiresome winter rains are pumping new life into the streams and rivers and building up reserves in the deep chalk to keep them bubbling through the hot dry days of summer. For the full clear flow, that brings the salmon in good condition to spawn below the drooping footbridge and scours the debris of the year from the bright gravel, has fallen on the chalk downs long, long before it runs sparkling past the watermeadows where the jewelled trout lie among the starwort and water buttercups.

There may be snow this year, bringing a new landscape of silent white or bright sunny mornings, when the world is hung with glittering crystals of frost and it is worth any effort to see the rich brown reed mace glowing against the cold blue ice. There will be wild geese on the flooded meadows and flocks of lapwings over the dead elm trees on the rim of the hill.

There will be fieldfares like big blue thrushes and redwings, too, in the holly and ivy trees and the hedgerows will be noisy with flocks of long-tailed tits and marauding finches. If it is not too cold to sit out, it will be ideal for drawing the bones and sinews of the naked trees. The cold will make me paint more quickly than the drowsy bee-loud days of summer and, with luck, I may catch the freshness of the crisp winter air.

The pool will have silted up a little more than last year. It is alive and always changing. Nature will fill it in again in the course of time and return it to the condition in which I found it. This is sad in a way, but inevitable, and paradoxically it is the most vital part of its enchantment. For, although it will take much longer, the pool itself will bloom like the celandines that star its banks in the spring, and fade away again as they do.

It does not matter, for, so long as man does not seal it in a concrete and brick tomb, it will be beautiful at every stage. What matters is to have been there when it bloomed and to have seen it in the sunshine and in the rain.

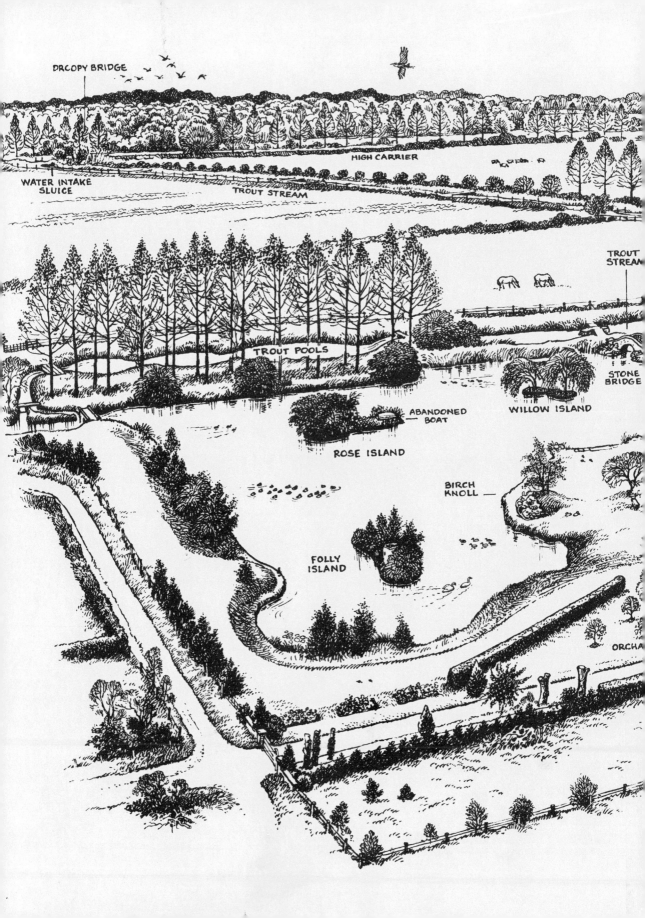

DRCOPY BRIDGE

WATER INTAKE
SLUICE

HIGH CARRIER

TROUT STREAM

TROUT
STREAM

TROUT POOLS

WILLOW ISLAND

STONE
BRIDGE

ABANDONED
BOAT

ROSE ISLAND

BIRCH
KNOLL

FOLLY
ISLAND

ORCHA